The C
Death Star

The Paleophysics of
the Great Pyramid and
the Military Complex at Giza

Joseph P. Farrell

The Lost Science Series:
The Anti-Gravity Handbook
Anti-Gravity & the World Grid
Anti-Gravity & the Unified Field
The Free-Energy Device Handbook
The Energy Grid
The Bridge to Infinity
The Harmonic Conquest of Space
The Fantastic Inventions of Nikola Tesla
Vimana Aircraft of Ancient India & Atlantis
The Time Travel Handbook
Ether Technology
The Tesla Papers
Lost Science
Secrets of Cold War Technology
Man-Made UFOs: 1944-1994
Technology of the Gods
Atlantis and the Power System of the Gods

The Lost Cities Series:
Lost Cities of Atlantis, Ancient Europe & the Mediterranean
Lost Cities of North & Central America
Lost Cities & Ancient Mysteries of South America
Lost Cities of Ancient Lemuria & the Pacific
Lost Cities & Ancient Mysteries of Africa & Arabia
Lost Cities of China, Central Asia & India

The Mystic Traveller Series:
In Secret Mongolia by Henning Haslund (1934)
Men & Gods In Mongolia by Henning Haslund (1935)
In Secret Tibet by Theodore Illion (1937)
Darkness Over Tibet by Theodore Illion (1938)
Danger My Ally by F.A. Mitchell-Hedges (1954)
Mystery Cities of the Maya by Thomas Gann (1925)
In Quest of Lost Worlds by Byron de Prorok (1937)

Write for our free catalog of exciting books and tapes.

The Giza Death Star

by

Joseph P. Farrell

Adventures
Unlimited
Press

The Giza Death Star
by Joseph P. Farrell

ISBN 0-932813-38-0

Printed in the United States of America

First printing June 2001

Published by
Adventures Unlimited Press
One Adventure Place
Kempton, Illinois 60946 USA
auphq@frontiernet.net
www.adventuresunlimitedpress.com
www.adventuresunlimited.nl

9 8 7 6 5 4

For

Tracy Scott Fisher
And
Peggy Hill,
Who Listened, Believed, and Encouraged

The Giza Death Star

The Paleophysics of the Great Pyramid and the Military Complex at Giza

Table of Contents

Table of Contents

Table of Contents

Preface

"Sir, we're all agreed your theory is crazy. The question that divides us is, whether it is crazy enough."
Nils Bohr to Werner Heisenberg

Even now, at the conclusion of this work, as I sit to write the preface to it, I am confident of partial failure.

When I first completed the draft of this book, little did I know that I would be embarking on a journey of discoveries more often than not frightening and terrible in their implications. The odyssey was initiated by my guess that the Great Pyramid coupled and oscillated gravitational and electromagnetic energy with acoustic energy and a superluminal wave form called a "pilot" or "scalar" wave. In that draft, I predicted that some redundant harmonics of Planck's constant, i.e., near-whole number multiples of it, would be discovered in the very areas where I hypothesized such engineering and coupling would occur: the Grand Gallery, the Antechamber, the King's Chamber, and the Queen's Chamber.

Shortly after completing the book and rushing it off to a publisher, I sat down with a calculator and began crunching numbers, and within two hours had found three close multiples of Planck's constant to within one decimal place. Excited by this discovery and my curiosity piqued, I spent the better part of two weeks crunching more numbers, and to my horrified delight, found not only the original three but *several* such redundancies right where I had originally hypothesized they would be. And not only was Planck's constant found, but the other "Planck units": the Planck length and mass, so essential to the Grand Unified Theories and String Theory of modern theoretical physics.

These results intrigued me, not only because they appeared to corroborate my hypothesis that the Great Pyramid was some paleoancient weapon of mass destruction employing an extremely sophisticated unified physics, but also because the *basis* of that

i

unification appeared to lie not in any theoretical model, but in the engineering itself.

It is rare for an author to be grateful that a publisher rejects a manuscript, but in this instance, I was appreciative because it allowed me the opportunity to incorporate my subsequent researches into the manuscript, which is presented here.

All that said, I remain confident of at least partial failure.

How does one adequately describe such a complex hypothesis about such a complex and baffling subject as the Great Pyramid? How does one describe adequately the most mysterious object on earth? Or the other mysterious objects around it? Or the mysterious geometric and astronomical arrangement in which they are all placed? And what is a large statue of a half-man half lion doing in the middle of it all, slouching to the east, gazing perpetually into space and time. And why is the Sphinx known in the Arabic tradition as "the father of terrors?"

But I am confident of failure for yet another reason. How does one survey, with adequate technical detail, such a complex hypothesis, and yet do so for a general audience? So I must warn the reader, here and now, that this is not a work for the faint hearted. It does not proceed in sound bites. It does not shun technical jargon, for any attempt to understand the Great Pyramid – on *any* hypothesis – that presumes to do so with sound bites or without technical jargon is doomed.

So what does the technical jargon of the weapon hypothesis have to say? It says that the Great Pyramid was a phase conjugate mirror and howitzer, utilizing Bohm's "pilot wave" as a superluminal carrier wave to accelerate cohered electromagnetic and gravito-acoustic waves to a target via harmonic interferometry. That rather tangled idea leads to a set of putative principles of its engineering. Since so many of its dimensional measures appear to be harmonically resonant to each other, the Pyramid, as a coupled harmonic oscillator, seems constructed of several oscillators nested within the structure in such a fashion as to suggest a set of feedback loops being used to amplify that oscillated energy.

That being said, the technical jargon is there not for a technical reason, but for a "mystical" one, as it were. It is there, like the Pyramid itself, to initiate into a mystery. And initiation always implies some degree of mental retraining, of modifying existing concepts by rearranging them in new relationships, and of reexamining and revitalizing older "outmoded" notions. This was, accordingly, a difficult book to arrange or to "sequence" properly.

For example, which should be treated first? The principles and outlines of modern theoretical physics and some of its more esoteric pursuits? Or the principles of the ancient "paleophysics" embodied in obscure ancient texts with ties to Egypt? Should one lay it out like a mathematics, physics, or engineering textbook, with every term neatly defined, and in its precise place, and thereby risk losing the reader in a blizzard of clarity of equations and diagrams? Or should one simply lay the terms out there and peal off the onion skins of meaning with ever more precise definitions, and thereby run the risk of losing the reader in the transparency of confusion?

I chose a middle course, and consequently, chapter two, "An Archaeology of Mass Destruction," and chapter three, "The Paleography of Paleophysics" simply discuss the evidences for a sophisticated and weaponized ancient technology, and the more abstruse notions of physics to be found in ancient texts, without much definition or clarification of terms. A certain degree of familiarity with contemporary physics is required, particularly with String theory and more importantly, with the much more satisfying Plasma Cosmology of Swedish physicist Hannes Alfvén.

In chapter four, *"De Physica Esoterica"* terms receive greater definition – for a more general audience - through a discussion of some of the stranger areas of contemporary theoretical physics, including some of the more startling and recent developments. Chapter five outlines in very cursory fashion the almost infinite and infinitely astonishing mathematical and physical properties of the Pyramid. Chapter six surveys Christopher Dunn's crucially important "machine hypothesis". Chapter Seven, "Planck in the Pyramid," surveys the various harmonic multiples of the Planck numbers found in various dimensional measures of the structure. In chapter eight, the whole is gathered into an exploration of the possible ways the Pyramid functioned as a weapons system of mass destruction. I have thus aimed for a "cumulative effect" that only begins to make sense, perhaps, in this chapter. Finally, in chapter nine, I explore various speculations of the type of society that could have built such an awesome and dreadful weapon, and then actually apparently used it.

I have chosen the title, *The Giza Death Star*, deliberately. At one level, it is an obvious resonance to Dunn's magisterial look at the machine hypothesis, *The Giza Power Plant*. At still another level, it is meant to conjure the well-known stellar alignments of the Pyramid to Orion and Sirius and the zodiacal associations with death that these constellations held for the Egyptians and for subsequent cultures. At a still deeper level, the title resonates to the appropriate images that Hollywood has conjured: planet busting moon-sized "Death Stars" from *Star Wars* or city-cracking electromagnetic pulse weapons from *Independence Day*.

And that is precisely the image I *wish* the title to conjure, for if the physics embodied in the ancient texts is what I think it is, and if its purpose was what the ancient texts suggest it was, then it was indeed a potential planet buster without parallel in the annals of military engineering. The physics embodied in it seems to touch upon every possible well-known aspect of contemporary theory, such as Bell's non-locality theorem and the various quantum states of the electron, to more esoteric subjects such as phase conjugation, non-equilibrium thermodynamics, harmonic oscillators, coherence, stealth technology, the Philadelphia Experiment, Montauk and all the other "fringe" areas of "pseudo-science" that squat like black beasts in the bowels of secret government research bunkers and prowl through the rumor mills of "alternative research" journals and books.

But at the most personal level, it reflects a personal fascination I have held with the structure since its deeply mysterious properties first became known to me. Unlike most people, however, I never experienced a sense of well-being when contemplating the Great

Pyramid. I have never felt anything "good" about it. I long held the notion that the architecture of Giza as a whole had about it the disquieting feel of a military compound. If one factored out the Sphinx for the purpose of a physical comparison, the resemblance between it and a modern military phased array radar seemed all too palpable.

So when I first read the texts reproduced by Zechariah Sitchin in his *The Wars of Gods and Men* I experienced that sickly sense that my worst fears might just be true. And when I read Dunn's work, those fears were confirmed, and I resolved to discover what principles, if any, might have made it work. What follows is therefore not a completed theoretical model, replete with equations and schematics and diagrams and so on. It is more like a preliminary field report, outlining research that remains to be done...

...or is perhaps already *being* done. I come away from this project with a feeling of profound disquiet that someone, somewhere, has been doing the type of "paleophysical" research this book implies. The litany of experimental and theoretical physicists involved with such esoteric research is known to anyone familiar with the literature. The names of Thomas Townsend Brown, Hal Puthoff, Oppenheimer, Tesla, Sagnac, von Neumann, DiPalma, Philo Farnsworth and many others fill the pages of literature that would never be found in most physics departments or laboratories, and yet, their preoccupations are there for all to see who would bother to look.

And that is the most disquieting result of all. The notion that there has existed – for millennia perhaps – an underground or esoteric tradition of research alongside the exoteric science one learns about in universities is a radical one. It implies the existence of a body of scientific knowledge that has been deliberately manipulated and suppressed. Indeed, one of the most disquieting features of Pyramid research recently has been the suppression of the findings of expeditions. And of course, one does not really know the extent of various governments' satellite and radar tomography imagining of the site. And the average researcher is certainly not privy to the still classified results of those studies.

In any case, if anything about *The Giza Death Star* and its weaponized physics herein presented is even approximately close to the actual truth, then we are indeed on the very cusp of a paradigm shift of earth-shattering geopolitical consequences. And for that reason, I hope that everything I say in this book is *not* true, and that my failure is total.

Joseph P. Farrell
Tulsa, Oklahoma 2001

Grand Gallery.

PART ONE:
PALEOPHYSICS

I.

Introduction:
Arcanum Organon

"You shall understand (that which perhaps you will scarce think credible) that about three thousand years ago, or somewhat more, the navigation of the world (especially for remote voyages) was greater than at this day."
Sir Francis Bacon, <u>The New Atlantis</u>[1]

The Great Pyramid is the most studied and surveyed building in the world, and for good reason. It is the largest, and most mysterious, human monument on the earth, an *arcanum organon*, a strange, arcane work. No other structure has so engaged the imagination nor evaded the efforts of scholars to explain. First, while it remains true that great strides in pyramidology have been made by those outside the myopic box of the paradigms of "orthodox history" and "Egyptology" – by engineers, physicists, geologists, astronomers, investigative journalists, or even fundamentalists of every stripe from "Christian" to "New Age" – in the final analysis each perspective sees mostly what it has been trained to see. Second, all these approaches seem to see in the Great Pyramid something beautiful and benignly wonderful, a triumph of geometry, astronomy, physics and engineering of some past and glorious golden age speaking its pacifistic wisdom in the mute silence of its stones down to our own time.

Many have stood at the base of that massive structure and marveled, with an understandable and overwhelming awe, at the civilization that could have built such a tomb, or observatory, or machine, or "prophecy in stone." I stand with them, looking up in awe and wonder, at the massive structure, and shudder at any civilization that could have built such a weapon, such a monument

---

[1] Sir Francis Bacon, *The Advancement of Learning and the New Atlantis* (Oxford: Oxford University Press, 1966), p. 271.

1

to the perversity, of mass destruction. So in that sense, yes, I agree with the "prophecy in stone" hypothesis, for that civilization was far too much like our own: capable of technological wonders, capable of mass destruction, and like our own, in almost complete moral decay. In that most profound sense, the Great Pyramid is a prophecy, and a warning.

This study is therefore a radical departure from previous attempts to explain the structures at Giza and their ultimate purpose. I do agree with some aspects of previous explanations. For example, I assume:

(1) that the "orthodox historiography" and conventional "Egyptological" explanations are simply incorrect in the extreme, and therefore:

(a) that the structures are the traces of some "paleoancient" Very High Civilization;[2]

(b) that humanity is therefore of far greater antiquity than is assumed by orthodox paradigms of history;

(c) that the structures embody a physics and technology of a civilization at least as advanced as our own if not much more so.

(d) that insofar as the current study agrees with orthodox Egyptology's religious interpretation of the function of the Giza structures, it maintains that those interpretations were *not* original to the civilization that built them, but were forced upon them subsequently by the legacy civilization that came ultimately to occupy them, namely ancient Egypt. Even then, as will be seen, these interpretations are partly the result of circumstances, and partly the result of a concerted effort and covert design on the part of that Paleoancient Very High Civilization to preserve the physical, geometric, and mathematical knowledge - and therefore the power - embodied in the structures.

---

[2] I use the rather redundant term "paleoancient" simply to designate a Civilization that once existed long before the "ancient civilizations" of standardized cultural history.

(2) that the various celestial, solar, and terrestrial relationships
embodied in the Pyramid are really there, and thus that the
Pyramid's subsidiary functions were indeed those of an
astronomic and temporal observatory and "time capsule."
But these were not the ultimate functions nor intentions of
its original builders. They did not intend to convey some
message or "wisdom" or "prophecy" to some future age, at
least, not initially. The ultimate purpose of these complex
mathematical and physical embeddings was far from the
benign astrological or prophetic vagaries later civilizations
and interpreters would place upon them. They were,
moreover, required by the kind of physics that I believe
made the Giza Death Star possible. Reconstruction of at
least parts of that physics is the task of this study.

Insofar then as I *disagree* with other non-orthodox interpretations,
I disagree only in that the technology of the civilization that built
Giza, and the use to which it was ultimately put, was far from
benign.

But before examining all that, we should first dispense with the
obvious absurdity, the sheer idiocy, of the standard explanation of
Egyptology that these massive structures – especially the Great
Pyramid and its almost infinitely anomalous engineering and
mathematical properties – are pharaonic tombs. No one is more
eloquent, impassioned, nor as on target in summarizing the "tomb"
hypothesis and its manifestly nonsensical nature than is Peter
Lemesurier, himself an exponent of the Pyramid-as-prophecy or
"time capsule" hypothesis.

> But how – and why (was it built)? The logic of the thing seems to defy
> all analysis.
> And so the historians.... Knowing precisely nothing about the
> project's origin...have naturally fallen back on a process of wild
> extrapolation from their only slightly less scratchy knowledge of later
> dynastic times. The Egyptians, it has been established, were obsessed
> with death and immortality, with the embalming of the dead, with
> preparations for life in the nether-world. Therefore the Great Pyramid
> Project represents the same obsession magnified to the Nth degree.
> And so the scene described for us is a kind of gothic melodrama

3

unequalled in its sheer antediluvian lunacy. The megalomaniac pharaoh Cheops, brooding over the fate of his own eternal soul, decides to throw his kingdom's entire resources into a colossal real-estate project designed purely to humour his own necromantic illusions of immortality. To satisfy this man's mere superstitious whim, thousands of slaves toil day after day to drag gigantic blocks of masonry up mighty ramps with the aid of nothing better than primitive sleds, levers, ropes and rollers. Overseers drawn from the serried ranks of Hollywood extras bark crude orders, wave cruder charts....

And the result? The Great Pyramid – a building so perfect and yet so enormous that its construction would tax the skills and resources even of today's technology almost to the breaking point....[3]

And not only would it tax the skills and resources of our technology, but suffice it to say that a project of this scope would literally "tax" the economies of today's most powerful nations.

So what then does one make of the "pharaonic tomb" hypothesis still found in every college textbook of the western world?

The sober truth is, of course, that no historian has yet advanced any explanation of the Great Pyramid's construction that is at all convincing. Nobody alive today knows for certain how the Pyramid was erected, how long it was in the building, how its near perfect alignments were achieved before the invention of the compass, or how its outer casing was joined and polished with such unsurpassed accuracy. Nor have historians produced any convincing theory as to why such an enormous undertaking, combined with such incredible accuracy, should have been deemed necessary for the construction of a mere tomb and funery monument to a dead king who in any case apparently never occupied it.[4]

In view of all the vast mathematics and physics embodied in this engineering feat, "We have no choice but to look at the questions of how and why afresh. The accepted answers just will no longer do."[5] And looking at such problems afresh implies that one look at the persistent paleographic traditions of all the world's oldest

---

[3] Peter Lemesurier, *The Great Pyramid Decoded* (Avon Books: 1977), pp. 8-9.

[4] Ibid., p. 6.

[5] Ibid., p. 11.

civilizations that imply a "cataclysmic destruction of an earlier world whose knowledge and technical achievements were far in advance of anything so far attributed by history to so-called early man, and may have indeed rivaled or even surpassed our own."[6]

So one is left with the three basic implied assumptions of any non-orthodox explanation of the structure:

(1) It was built by humans[7] of a paleoancient Very high Civilization for a purpose which can best be ascertained:
  (a) by a careful analysis of relevant ancient texts, and
  (b) by comparison of those texts with an analysis of the structure itself on the basis of the latest physical theories and solid scientific hypothesis and conjecture;

(2) the human civilization that built the Pyramid was of a particular type possibly capable of interplanetary travel; and,

(3) Subsequent civilization, representing a decline in scientific and technological sophistication from the civilization which built it, placed interpretations upon it in the most technical language available to them, namely, the religio-astrological explanations common to ancient priest-craft.

A. An Old Turkish Map

The idea that there was once a paleoancient Very High Civilization of great scientific and technological sophistication is central to the thesis that the Great Pyramid was some sort of weapon and that some of the older structures of the Giza compound constituted a military complex. Most researchers, except of course the Egyptologists and historians in the universities, are agreed on the existence of some such paleoancient Very high Civilization. The lineaments and traces of this civilization are there for all to see, in almost every part of the globe, in the most ancient traditions of aboriginal tribes, in the

---

[6] Lemesurier, op. cit., p. 12.
[7] Humans, not aliens, in distinction from Sitchin.

most abstruse esoteric texts, in the most massive monuments, and in the puzzling anomalies that defy our most cherished academic notions of history. But what *are* these treasured notions?

> At some risk of over-simplification, the academic consensus is broadly:
> - Civilization first developed in the Fertile Crescent of the Middle East.
> - This development began after 4000 BC, and culminated in the emergence of the earliest true civilizations (Sumer and Egypt) around 3000 BC, soon followed by the Indus Valley and China.
> - About 1500 years later, civilization took off spontaneously and independently in the Americas.
> - Since 3000 BC in the Old World (and about 1500 BC in the New) civilization has steadily "evolved" in the direction of ever more refined, complex and productive forms.
> - In consequence, and particularly by comparison with ourselves, all ancient civilizations (and all their works) are to be understood as essentially primitive (the Sumerian astronomers regarded the heavens with unscientific awe, and even the pyramids of Egypt were built by 'technological primitives').[8]

One of the anomalies that threw all this into a cocked hat was the discovery of the map of the Turkish admiral Piri Reis in the 16th century. Drawn from earlier exemplars, this map was remarkable in that it detailed the southern Atlantic coastline of South America, but also the coastline of *Antarctica*. Perhaps that would not be remarkable in and of itself, save for the amazing accuracy the map displays of coastline submerged beneath tons of ice, and only recently discovered in the technologically sophisticated 19th and 20th centuries.[9] Even this, perhaps, would not be significant in and of itself, save for the fact that the Turkish admiral's map is not the *only* such map in existence detailing geographical knowledge of the New World, *long* before its "discovery" by Columbus.

---

[8] Graham Hancock, *Fingerprints of the Gods* (New York: Three Rivers Press, 1995), pp. 11-12.

[9] For the best, and the classic, treatment of this strange anomaly, cf. Charles Hapgood, *Maps of the Ancient Sea Kings* (Kempton, Illinois: Adventures Unlimited Press).

Author and investigative journalist Graham Hancock puts the case for these cartographic anomalies very succinctly:

> It would be futile to speculate further than Hapgood has already done as to what 'underground stream' could have carried and preserved such knowledge through the ages, transmitting fragments of it from culture to culture and from epoch to epoch. Whatever the mechanism, the fact is that a number of other cartographers seem to have been privy to the same curious secrets.
>
> Is it possible that all these mapmakers could have partaken, perhaps unknowingly, in the bountiful scientific legacy of a vanished civilization?[10]

But what civilization? And more importantly, what sort of "bountiful scientific legacy" did it actually posses?

B. *"Curiouser and Curiouser":*
Anomalous Monuments and Artifacts

Whether or not his interpretations were valid, whether or not his argumentation was sound, whether or not his hypothesis exceeded the evidence adduced for it, Erik von Däniken's *Chariots of the Gods* did serve to illuminate the problem of ancient anomalous monoliths clearly. From the massive earthen pyramids in China and the vitrified cities of the Indus Valley civilization, to the massive structures of Chichen Itza, Stonehenge, Easter Island, and a host of places in Central and South America, the world is dotted with anomalous, inexplicable structures and artifacts of by-gone civilizations.

But perhaps the strangest anomaly of all is the existence of such civilizations. Why, out of nowhere, did humanity leap from hunter-gathering tribal societies to the glories of Egypt, Sumer, the Incas, the Olmecs, and the Chinese? The records, at least as normally interpreted, provide no clues.

Standard history is loathe to consider seriously what most of those civilizations say about themselves, and "orthodox

---

[10] Hancock, op. cit., p. 13.

archaeology" and anthropology have committed too much faith in the extension of the evolutionary paradigm to the history of human culture to give a second thought to what those ancient monoliths and artifacts might actually be saying. Despite these inhibitions, the records, monoliths, and artifacts speak clearly enough: these civilizations, without exception, all considered themselves to be the legacies of an older, and much more sophisticated culture. That culture, located in the mists of the distant past, was a golden age, when "gods" mingled with men and directed their affairs, when great technological wonders were wrought, when basic morality and decency were in collapse, when tremendous wars were fought with horrific weapons, and when a cataclysm – the judgement of the "gods" or of God – overtook mankind.

But what are some of these artifacts and monoliths and records? Graham Hancock has artfully catalogued many of them in his *Fingerprints of the Gods*. While it is impossible to enumerate them all here, nevertheless some of them deserve mention, as some of the outlines of that paleoancient Very High Civilization and its technological sophistication emerge. And there is one artifact not mentioned by Hancock that is also worth looking into, by way of introduction into his own astonishing catalogue of mysteries.

That artifact is the Mitchell-Hedges Crystal Skull. This solid quartz crystal skull, discovered by the daughter of the British archaeologist F.A. Mitchel-Hedges while both were on an archaeological dig in the British Honduras in the 1920s, is like the Great Pyramid at Giza one of the world's most ancient, perfect, and anomalous artifacts. It was discovered at Lubaantum, a massive Mayan center built sometime around the 8th or 9th centuries A.D.[11]

But what is so anomalous about a solid quartz crystal model of a human skull, beyond its recovery at an ancient Mayan archaeological site?

---

[11] Alice Bryan, Phyllis Galde, *The Message of the Crystal Skull: From Atlantis to the New Age* (St. Paul, Minnesota: Llewellyn Publications, 1991), p. 18.

In the 1960s, Anna Mitchell-Hedges brought the skull to the attention of the noted art conservator Frank Dorland, who was instrumental in the authentication of the Icon of the Black Virgin of Kazan. Unlike most works of art, Dorland immediately recognized one problem: "there was no legend, no myth, no record, no reference to go on."[12]   In other words, the Mayans were unusually quiet about this most remarkable work of art, if indeed, that is what it was intended to be, for the skull began to divulge some very peculiar properties to the mystified Dorland.

> Underneath his microscope, Dorland began to discover incredible optics of a very sophisticated nature. Halfway back in the roof of the mouth of the Skull there is a broad, flat plane similar to a 45 degree prism. This surface can direct light from beneath the Skull into the eye sockets
>
> If the Skull were placed on a stone altar having a concealed interior firebox and a light hole up through the stone to where the Skull was sitting, the flickering flames could be reproduced visually as being alive in the eye sockets. There is also a thin ribbon-like surface carved next to this flat plane that could act as a magnifying reading glass. Next to the 45 degree prism there is a natural ribbon prism. Extending through the more than six inches of solid quartz crystal, this channel is free from veils and inclusions. Print viewed through this is not only legible, but also undistorted and only slightly magnified.
>
> Behind the intentionally carved prism, there is a concave and convex surface that acts as a light gatherer to bounce the light to the 45 degree prism and out to the eye sockets. The back of the Skull is formed as a beautiful camera lens, gathering light anywhere from the rear and reflecting it into the eye sockets.[13]

The sophisticated optics are our primary interest here, as they bear directly on the possible technology that built the Pyramid. Up against such a daunting display of optics in what presumably was only an artifact of religious art, Dorland decided to take the skull to the Hewlett-Packard Laboratories in Santa Clara, California to be tested.[14]

---

[12] Ibid., p. 36.
[13] Bryant and Galde, op. cit., p. 36.
[14] Ibid., p. 42.

There the mysterious nature of the anomaly only grew. The technicians at Hewlett-Packard

> performed two significant tests on the Crystal Skull. Submerging it in a bath of index-matching benzyl alcohol and viewing it under a polarized light, they determined that it had been cut without regard to the axis and that it was a single crystal. The orientation of the X-Y axis and the "veils" revealed by the polarized light showed that the jawbone, now a completely separate piece, had originally come from the same piece of crystal.... One worker said, 'There is no way of proving its age....'
>
> Even among people familiar with crystal and its properties, it raised as many questions as were answered. The exquisite workmanship and high gloss of the finish cause it to appear brand new, but it was the consensus of the lab experts that, given a crystal of the same size, these foremost producers of crystal components in the world today could not possibly produce a skull of comparable quality.[15]

So there it is: a non-dateable solid quartz crystal human skull, with prisms and ribbons "carved" on the *inside* of the skull, which in turn is "carved" from a *single* piece of quartz with a lapidary skill not possessed today, found in an ancient Mayan site, for which no explanation exists coming from that society. But the mysteries of pre-Columbian Central and South America do not stop there.

A site that figured prominently both in von Däniken's and Hancock's study is Nazca, in southern Peru. There on the plain one may find enormous drawings of spiders, monkeys, and "hummingbirds", but they are so large that they can only be seen from the air. Of particular interest is the Nazca "spider":

> Recent research by Dr Phillis Pitluga, a senior astronomer with Chicago's Adler Planetarium, has demonstrated that the spider, like the Great Pyramids at Giza in Egypt... was designed as a terrestrial image of the constellation of Orion. Is it possible that the incorporation of a 'celestial plan' in ancient and mysterious monuments from many different parts of the world, and a particular focus on the three stars of Orion's Belt (represented at Nazca by the narrow waist of the spider),

---

[15] Ibid., pp. 42, 44.

could be parts of a global scientific legacy passed on by a lost civilization of very remote times?[16]

But the Nazca spider is not the only mysterious site in the "new" world. Further north, northeast of Mexico City, lies the ancient city of Teotihuacan.

Here, as in Giza, the astronomical knowledge springs from nowhere, fully-fledged. Hancock notes that Stansbury Hagar believes the street of the city to represent the Milky Way.

> Indeed Hagar went further than this, seeing the portrayal of specific planets and stars in many of the pyramids, mounds, and other structures that hovered like fixed satellites around the axis of the Street of the dead. His complete thesis was that Teotihuacan had been designed as a kind of 'map of heaven:' 'It reproduced on earth a supposed celestial plan of the sky-world where dwelt the deities and spirits of the dead.'[17]

Anxious to test this theory, an American engineer named Hugh Harleston Jr. journeyed to Teotihuacan. Surveying the grounds carefully, he came to an even more astonishing conclusion:

> What Harleston's investigations had shown was that a complex mathematical relationship appeared to exist among the principal structures lined up along the street of the dead (and indeed beyond it). This relationship suggested something extraordinary, namely, that Teotihuacan might originally have been designed as a precise scale-model of the solar system. At any rate, if the centre line of the Temple of Quetzacoatl were taken as denoting the position of the sun, markers laid out northwards from it along the axis of the Street of the dead seemed to indicate the correct orbital distances of the inner planets, the asteroid belt, Jupiter, Saturn (represented by the so-called 'Sun' Pyramid), Uranus (by the 'Moon' Pyramid), and Neptune and Pluto by as yet unexcavated mounds some kilometers farther north.[18]

But why is this unusual?

---

[16] Hancock, op. cit., plate 1 caption.

[17] Ibid., p. 166.

[18] Hancock, op. cit., p. 167, emphasis added.

> If these correlations were more than coincidental, then, at the very least, they indicated the presence at Teotihuacan of an advanced observational astronomy, one not surpassed by modern science until a relatively late date. Uranus remained unknown to our own astronomers until 1787, Neptune until 1846 and Pluto until 1930. Even the most conservative estimate of Teotiheacan's antiquity, by contrast, suggested that the principal ingredients of the site-plan... must date back at least to the time of Christ. No known civilization of that epoch, either in the Old World or in the New, is supposed to have had any knowledge at all of the outer planets – let alone to have possessed accurate information concerning their orbital distances from each other and from the sun.[19]

Perhaps, given the sophistication of this knowledge at Giza and Teotihuacan, it is more accurate to speak of the rediscoveries of modern science.

But there is more by way of anomalous parallels between the Old and New Worlds' ancient civilizations.

C. An Algorithmic Language, Cuneiform Mathematics, and the Curious Geometry of Ancient Hebrew

Close to Lake Tiahuanaco in South America there is an ancient Indian tribe known as the Aymara, who speak a language "regarded by some specialists as the oldest in the world."[20] This language has some peculiar properties indeed:

> In the 1980s Ivan Guzman de Rojas, a Bolivian computer scientist, accidentally demonstrated that Aymara might be not only very ancient but, significantly, that it might be a 'made-up' language – something deliberately and skilfully (sic.) designed. Of particular note was the seemingly artificial character of its syntax, which was rigidly structured and unambiguous to an extent thought inconceivable in normal 'organic' speech. This synthetic and highly organized structure meant that Aymara could easily be transformed into a computer algorithm to be used to translate one language into another: 'The Aymara Algorithm

---

[19] Ibid., emphasis in the original.
[20] Hancock, op. cit., p. 91.

12

is used as a bridge language. The language of an original document is translated into Aymara and then into any number of other languages.'[21]

A designed algorithmic human language? Curious, indeed, but surely only coincidence.

But this is not the case. A similar mathematical property also exists in the ancient cuneiform of Sumeria, thousands of miles away.

> Mathematicians, especially those dealing with graph theory - the study of points joined by lines – are familiar with the Ramsey Graph Theory, named for Frank P. Ramsey, a British mathematician who, in a paper read to the London Mathematical society in 1928, suggested a method for calculating the number of various ways in which points can be connected and the shapes resulting therefrom.... The Theory offered by Ramsey made it possible to show, for example, that when six points representing six people are joined by either red lines (connecting any two who know each other) or blue lines (connecting any two who are strangers), the result will always be either a red or a blue triangle. The results of calculating the possibilities for joining (or not joining) points can best be illustrated by some examples. Underlying the resulting graphics (i.e., shapes) are the so-called Ramsey Numbers, which can be converted to graphs connecting a certain number of dots. I find that this results in dozens of "graphs" whose similarity to the Mesopotamian cuneiform signs is undeniable."[22]

Yet another Semitic language has oddly, and hugely, mathematical properties. The mathematician Stan Tenen has computer modeled the characters of the Hebrew alphabet, demonstrating not only rotational symmetries and torus knots, but that

> An extraordinary and unexpected geometric metaphor (exists) in the letter sequence of the Hebrew text of Genesis that underlies and is held in common by the spiritual traditions of the ancient world. This

---

[21] Ibid., emphasis in the original.
[22] Zecharia Sitchin, *Genesis Revisited* (Avon, 1990), pp. 225-227. The charts of comparison that Sitchin produces are provocative and constitute significant evidence of a paleoancient Very High Civilization.

metaphor models embryonic growth and self-organization. It applies to whole systems, including those as seemingly diverse as meditational practices and the mathematics fundamental to physics and cosmology... (demonstrating) that the relationship between physical theory and consciousness, expressed in explicit geometric metaphor, was understood and developed several thousand years ago.[23]

Mathematically and geometrically modeled languages that only now are being understood because of the invention of the calculus and modern computers suggests that a similar technology once existed on earth.

And that, of course, only increases the mystery.

D. Strange Religious Parallels: Virachocha and Osiris

It is reasonable to assume that if there are such strong scientific correlations between the ancient civilizations of the Old and New Worlds, that there might also be correlations in their religions as well. The ancient Incan Creator God, Viracocha, for example, possesses some interesting parallels with the Egyptian myth of Osiris.

Although there are huge difference between the traditions it is bizarre hat Osiris in Egypt and Thunapa-Viracocha in South America should have had all of the following points in common:
- both were great civilizers;
- both were conspired against;
- both were struck down;
- both were sealed inside a container or vessel of some kind;
- both were then cast into a river;
- both eventually reached the sea.

Are such parallels to be dismissed as coincidences or could there be some underlying connection?[24]

The difficulty in establishing this connection is almost insurmountable.

---

[23] www.meru.org. May 31, 2000, p. 1.
[24] Hancock, op. cit., p. 69.

The Olmec civilization in ancient Mexico is one example of this problem. Olmec carvings depict a racial class not indigenous to the area at all.[25] One might refer to this as the "Synoptic Problem" of ancient civilizations: why do so many emerge in such widely disparate places and so fully formed? Hancock's answer to this question is intriguing:

> It occurred to me that one plausible explanation might lie in a variant of the 'hypothetical third party' theory originally put forward by a number of leading Egyptologists to explain one of the great puzzles of Egyptian history and chronology.
>
> The archaeological evidence suggested that rather than developing slowly and painfully, as is normal with human societies, the civilization of Ancient Egypt, like that of the Olmecs, emerged all at once and fully formed. Indeed, the period of transition from primitive to advanced society appears to have been so short that it makes no kind of historical sense. Technological skills that should have taken hundreds of even thousands of years to evolve were brought into use almost overnight – and with no apparent antecedents whatever.[26]

The ultimate question boils down to "what was that 'common but exceedingly ancient source?'" It makes no rational sense to "argue" merely that civilization "took off" much later in Mexico than in Egypt and Mesopotamia.[27] Indeed, that is not an argument at all, merely avoidance of the issue.

But having pointed out the fallaciousness of that line of reasoning, Hancock then proceeds to elaborate a different hypothesis, and it is worth quoting him extensively at this point:

> Finally, let us turn again to Egypt of the Pyramid Age and the privileged status of pharaoh, which enabled him to circumvent the trials of the underworld and to be reborn as a star. Ritual incantations were part of the process. Equally important was a mysterious ceremony known as 'the opening of the mouth', always conducted after the death of the pharaoh and believed by archaeologists to date back to pre-dynastic times. The high priest and four assistants participated,

---

[25] Ibid., plates 25-28.

[26] Ibid., p. 135.

[27] Ibid., p. 137.

wielding the *peshenkhef*, a ceremonial cutting instrument. This was used 'to open the mouth' of the deceased God-King, an action thought necessary to ensure his resurrection in the heavens. Surviving reliefs and vignettes showing this ceremony leave no doubt that the mummified corpse was struck a hard physical blow with the *peshenkhef*. In addition, evidence has recently emerged which indicates that one of the chambers within the great Pyramid at Giza may have served as the location for this ceremony.

All this finds a strange, distorted twin in Mexico. We have seen the prevalence of human sacrifice there in pre-conquest times. Is it coincidental that the sacrificial venue was a pyramid, that the ceremony was conducted by a high priest and four assistants, that a cutting instrument, the sacrificial knife, was used to strike a hard physical blow to the body of the victim, and that the victim's soul was believed to ascend directly to the heavens, sidestepping the perils of the underworld....

Could it be, therefore, that what confronts us here, in widely separated geographical areas, and at different periods of history, is not just a series of startling coincidences but some faint and garbled common memory originating in the most distant antiquity? It doesn't seem that the Egyptian ceremony of the opening of he mouth influenced directly the Mexican ceremony of the same name (or vice versa for that matter). The fundamental differences between the two cases rule that out. What does seem possible, however, is that their similarities may be the remnants of a shared legacy received from a common ancestor. The peoples of Central America did one thing with that legacy and the Egyptians another, but some common symbolism and nomenclature was retained by both.[28]

We are now in a position to trace the outlines of the paleoancient Very High Civilization. There is one important distinction that must be pointed out. In Hancock's version of this thesis, Egypt and Mesopotamia are the legacies of that civilization, and Giza itself, in that model, is to some extent a part of that legacy, a part of Egypt. But in the version I am proposing, the

---

[28] Hancock, op. cit., pp. 142-143. The fact that there is some evidence that the interior of the Great Pyramid may have been used by the Egyptians in the 'opening of the mouth' ceremony does not invalidate the *earlier* use of it as a weapon by the paleoancient Very High Civilization.

Great Pyramid itself is the monument, the artifact, perhaps the only one remaining, of that paleoancient Very High Civilization.

What are the common outlines of this Very High Civilization that emerge from this cursory survey? If we assume that these artifacts are all the result of having sprung from some common source, then those outlines become rather clear:

- It was a civilization preoccupied with death, or rather, death was a theme that the legacy civilizations retained from the more ancient precursor, and this theme may have been distorted in transmission.
- It was also a civilization possessed, apparently, of sophisticated physical (including optical), astronomical, and mathematical science.
- It was also a civilization that possessed some form of advanced computational technology and the knowledge of algorithms that goes with it.
- The paleoancient Very High Civilization that built the Great Pyramid and perhaps some of the other monuments at Giza was based very likely in North Africa and in Mesopotamia, since the legacy civilizations spring up there first.
- The paleoancient Very High Civilization was possibly global in extent, a fact that may account for the "simultaneous" emergence of civilizations in Egypt and the Americas. This is not to say that there were not several societies *within* that global culture, for indeed, the records indicate that this culture ripped itself apart in a series of devastating wars, to be finally buried in a cataclysm of environmental disasters.
- A certain amount of evidence exists to suggest that one element or society within that culture was strongly possessed by an ideology that applied its technology and sophisticated science in the pursuit of vastly evil goals, perhaps attempting to dominate other societies within it. This would suggest that one society in

17

particular was rather more technologically advanced than the rest.[29]

- The legacy civilizations either intentionally garbled the religious and scientific knowledge of their precursor, or that blending was itself a legacy from that precursor, or the blending was perhaps a result of both processes. As we shall eventually see, the likelihood is that the blending is original to the paleoancient Very High Civilization, even though there is inevitable, and perhaps intentional, garbling in transmission.
- Finally, it was a civilization that was equally preoccupied with immortality, the afterlife, and time.

E. The Wrong Paradigm?

Hancock and Bauvall indicate that the "wrong paradigm" or "program" has been used when interpreting the ancient Egyptian Pyramid Texts. The attempt to understand the Pyramids, and the Pyramid Texts, in any other way than the way indicated by the religion of the Egyptians themselves is, for them, precisely the wrong paradigm.[30] But this appears to be too conventional, for neither the science, nor the technology, nor even the religion embedded in the Pyramid by its builders, was in any way conventional. But for all that, we shall discover in the next chapter that the Pyramid's builders' religion was based primarily in the physics, and not vice versa.

---

[29] From the Bible to the Aztec myths of Quetzocoatl, the image of a serpent bringing enlightenment, as well as the demand for blood, is persistent and pervasive.

[30] Cf. Graham Hancock, Robert Bauvall, *The Orion Mystery* (New York: Crown Publishers, 1994), pp. 73-80.

II. An Archaeology of Mass Destruction

From the sands of Egypt to the Indian subcontinent to the vitrified fortifications of northern Scotland, there are anomalous evidences that point not only to an enormous catastrophe that once beset the earth, but to a global war fought with weapons of mass destruction that may have triggered it. These anomalies form the scenery in which to view the workings and deployment of the Giza Death Star. They constitute compelling corroborative evidence that the Great Pyramid may once have been one of the weapons, perhaps *the* "Great Weapon" and "Great Affliction," whose deployment wrought such destruction and initiated such disasters.[1]

In this chapter we will sketch this scenery in very broad strokes. First, we will discover evidence of a paleoancient global war, fought with thermonuclear weapons and demonstrate that evidence exists that weaponry even more powerful and destructive may have been deployed. Then we will survey the very suggestive – and highly speculative hypothesis - that much ancient art and occult symbolism contains detailed schematics of very sophisticated technological devices, including, among other things, a three stage fission-fusion-fission hydrogen bomb. Finally, we will explore the evidence that suggests that certain governments and secret societies are actively and secretly pursuing the reconstruction of the science and technology that made the Giza Death Star possible.

A. Evidence of a Paleoancient War Utilizing Weapons of Mass Destruction

In an excellent study called *Technology of the Gods: the Incredible Sciences of the Ancients*, author David Hatcher Childress explores the evidence that exists for an "Ancient Atomic Warfare."[2] On February 16, 1947, the *New York Herald Tribune* ran an article about an unusual set of archaeological anomalies.

---

[1] Zechariah Sitchin, *The Wars of Gods and Men* (Avon, 1985), p. 174.

The article stated, rather matter of factly and without much reflection on its significance, that

> When the first atomic bomb exploded in New Mexico, the desert sand turned to fused green glass. This fact, according to the magazine *Free World*, has given certain archaeologists a turn. They have been digging in the ancient Euphrates Valley and have uncovered a layer of agrarian culture 8000 years old, and a layer of herdsman culture much older, and a still older caveman culture. Recently, they reached another layer...of fused green glass.

The next statement is the only indication that the newspaper gave of its real significance: "Think it over, brother."[3]

Of course, such fused glass can be caused by powerful lightening strikes during thunderstorms. There are, in fact, a class of archaeological and geological curiosities called fulgurites. These are branched, tubular globs of fused green glass caused by such lightening strikes.[4] The trouble is, such strikes do *not* explain whole areas of desert of a more or less uniformly and circularly distributed pattern of green glass. Many researches into the hypothesis of a paleoancient Very High Civilization have concluded that these blasts were caused by nuclear and thermonuclear detonations in a long-since fought atomic war.

Standard academic theory is quick to dismiss this idea, pointing out that meteors of sufficient size impacting the desert would generate sufficient mechanical energy to fuse the glass. The trouble with this view is that meteor impacts leave *craters*, whereas atomic explosions are normally air bursts whose tremendous heat fuses the silica of desert sand and dirt. And this is precisely what one encounters in the "green glass" layer: no craters, and vast areas of fused green glass.

---

[2] David Hatcher Childress, *Technology of the Gods: The Incredible Sciences of the Ancients* (Kempton, Illinois: Adventures Unlimited Press, 2000), pp. 211-257.

[3] Childress, op. cit., p. 211.

[4] Ibid., p. 212.

Researchers into the paleoancient Very High Civilization hypothesis point out that there were at least three known societies within this civilization:

- The Rama Empire, based in the Indian subcontinent;
- The "Osirian" society, based in North Africa and the Mediterranean; and,
- Atlantis, based either in the Atlantic or Antarctica.

Other variations of the hypothesis include "Lemuria", a lost continent connecting Indochina, Burma, and Malaysia with Australia, and "Mu", a lost civilization buried under the waters of the Pacific.

Four our purposes we will note that in most versions, the Osirian civilization that once occupied the Eastern Mediterranean basin and North Africa – and therefore Egypt – was, along with Rama, a relatively peaceful and benign society. In all versions of the hypothesis, Atlantis was the warlike society.

However, as noted in his book *The Wars of Gods and Men*, Zechariah Sitchin believes that the war which erupted between the various societies within this paleoancient global civilization was largely fought to destroy the "Great Weapon", the Giza Death Star, the Great Pyramid at Giza. I concur with that analysis. But this means that a reconstruction of the hypothesis is necessary, for it is evident that the society that built the Giza Death Star was hardly the peaceful and benign "Osirian" society. Accordingly, I believe that the "Atlantean" society extended itself to North Africa, and specifically to Egypt. The later civilization of Egypt itself was therefore most likely a blending of Atlantean and Orisirian elements. The war would have been fought, therefore, between elements of the Osirian society - based in the Eastern Mediterranean, Anatolia, the fertile crescent - and Rama, based in the Indian Subcontinent on the one hand, and the Antlantean civilization.

Supporting this hypothesis, one finds curious, if not horrifying, evidence that atomic weapons were deployed against Egypt, the Near East, and the Indian sub-continent. In the southwest Egyptian desert, near the borders of Libya, Chad, and Sudan, there is a vast

21

sea of such fused green glass known simply as "The Libyan Desert Glass."[5] In India, of course, whole cities have been "vitrified" by intense heat. Human skeletons lie in streets of stone, oftentimes holding hands with each other or engaged in normal human activities, as if suddenly burnt and fused with the molten stone.[6] And near modern Bombay, the most startling, and thought-provoking, evidence of all: The Lonar Crater. 2,154 meters in diameter, and

> Aged at less than 50,000 years old, (it) could be related to nuclear warfare of antiquity. No trace of any meteoric, etc., material has been found at the site or in the vicinity, while it is the world's only known 'impact' crater in basalt. Indications of great shock (from a pressure exceeding six hundred thousand atmospheres) and intense abrupt heat (indicated by basalt glass spherules) can be ascertained at the site.[7]

If the crater was caused by a weapon, then it may not have been by a nuclear weapon, since nuclear weapons are not *impact* weapons.

As Childress points out

> Vitrification of brick, rock and sand may have been caused by any number of high-tech means. New Zealand author Robin Collyns suggests in his book *Ancient Astronauts: A Time Reversal?* That there are five methods by which the ancients or "ancient astronauts" might have waged war on various societies on planet Earth. He outlines how these methods are again on the rise in modern society.
>
> The five methods are: plasma guns, fusion torches, holes punched in the ozone layer, manipulation of weather processes, and the release of immense energy such as an atomic blast....
>
> Explaining the plasma gun, Collyns says, "the plasma gun has already been developed experimentally for peaceful purposes: Ukrainian scientists from the Geotechnical Mechanics Institute have experimentally drilled tunnels in iron ore mines by using a plasmatron, i.e., a plasma gas jet which delivers a temperature of 6,000°C."

---

[5] Childress, op. cit., p. 214.
[6] Childress, *Vimana Aircraft of Ancient India and Atlantis* (Kempton, Illinois: Adventures Unlimited Press,), p. .
[7] Childress, *Technology of the Gods*, p. 239.

A plasma, in this case, is an electrified gas…

Collyns goes on to describe a fusion torch: "This is still another possible method of warfare used by spacemen, or ancient advanced civilizations on Earth. Perhaps the solar mirrors of antiquity really were fusion torches? The fusion torch is basically a further development of the plasma jet. In 1970 a theory to develop a fusion torch was presented at the New York aerospace science meeting by Drs. Bernard J. Eastlund and William C. Cough. The basic idea is to generate a fantastic heat of at least fifty million degrees Celsius which could be contained and controlled.[8]

In later chapters I outline my hypothesis that the nuclear energy of the hydrogen plasma in the Great Pyramid was coupled to a superluminal "pilot wave" along with acoustic and electromagnetic energy and guided via harmonic interferometry to a target. The result, I indicated, would be an enormous thermonuclear and nuclear reaction in the target, *regardless of the elements which composed the target.*

Finally, there is corroborating evidence in ancient Chinese legends that refer to a monstrous weapon known as a "yin-yang mirror." Collyns observes that "It is not technically impossible for the solar mirrors to have reflected light and heat (and electromagnetic?) radiation from a central radiant core, e.g., a plasma radiation energy source positioned in the center of a crystalline/metallic alloyed mirror, and held by a magnetic field."[9] As will be seen subsequently, the Great Pyramid was designed not only as a crystal and a mirror, but employed precisely such a plasma, drawing upon the energy of the sun, the solar system, and galaxy itself, as well as upon the energy that propels the sun: thermonuclear fusion.

So what, then do we have?

- Vitrified cities and fortifications, whose radioactive content, in the case of Indian sites, is matched only by the radiation of Hiroshima and Nagasaki.
- Large layers of fused green glass in Sumeria and Egypt.

---

[8] Childress, *Technology of the Gods,* pp. 224-225.
[9] Cited in Childress, *Technology of the Gods,* pp. 131-132.

23

- Ancient Chinese texts that refer to a super-weapon called a "yin-yang" mirror.
- Ancient Sanskrit and Hindu texts that refer to a super-weapon "Charged with all the power of the universe."
- Sumerian texts that refer to the Great Pyramid as "The Great Weapon" and "the Great Affliction."

B. Evidence of Sophisticated Paleoancient Engineering:
Allusions of Circuitry and Schematics in Ancient Art, Language,
and Pictographs

Ancient art, language and pictographs contain allusions and resemblances to sophisticated circuitry and other schematics. Moreover, occult systems such as Rosicrucianism, masonry, the Qabala and Alchemy, not to mention ancient classical systems such as Pythagoreanism, contain a great deal of sophistication in the display of complex, and apparently meaningless geometrical patterns. However this presents certain difficulties in terms of the examination of these artifacts for their possible significance in explication of paleophysics and its engineered applications.

- Any comparison of the pictographs and symbols of modern circuitry schematics and ancient art can only be aesthetic in nature, since one cannot determine on any *prima facie* or "first look" comparison that there is anything more than an analogy of form present in such comparisons. This is because the pictographs of modern circuitry schematics and that of the paleophysics – if it existed at all - may have been entirely different. We have no way of knowing for certain that such ancient art and artifacts were intended to encode or disguise such schematics.
- Secondly, any simple comparison does not account for the historical origins of our own schematic pictographs. Establishing the origin of each pictograph within our own schematic language is difficult. If the origins of such pictographs could be traceable to any esoteric or occult provenance, this would constitute corroboration of the

hypothesis that a certain physics and engineering was deliberately encoded in secret occult traditions. Moreover, if the person or persons originating the use of a certain pictograph within the modern schematic language could be documented to be a member or associate of members of such traditions, then this corroboration would be stronger.

However, the fact remains that the resemblances are there, and it must therefore be conceded that such diagrams might be construed as technological information.

1. Two Faces of Viracocha

In the ancient ruins at Lake Titicaca in Bolivia, the famous "Sun Gate" contains a mural, the central figure of which most scholars agree to be the Creator-God Viracocha.[10] The unusual feature of this mural, besides its astonishing workmanship and precise symmetry, is the ease with which it bears an analog to certain technical devices found only in the latter half of the twentieth century.

If one understands the mural to be a schematic, and retains the same rotational plane, the mural could just as likely be depicting a vacuum tube of some sort, complete with electrodes, plugs, and circuitry. More suggestive, however, is its over-all resemblance to a bomb, complete with fins, altimeter, firing circuitry, and implosion devices surrounding a critical mass of fissile material divided into three parts, with possible shields of neutron initiators (polonium and so on), according to the following schematic. In other words, on one view, one has here a very simple schematic for a workable three stage fission-fusion-fission bomb, complete with implosion detonator and surrounding casing of fissile material. An atomic bomb is needed to generate the necessary heat to initiate a hydrogen fission reaction, which in turn generates enough energy to cause further fission in the outer casing of the bomb.

---

[10] Zecharia Sitchin, *The Lost Realms, Book IV of The Earth Chronicles* (Avon, 1990), p. 218.

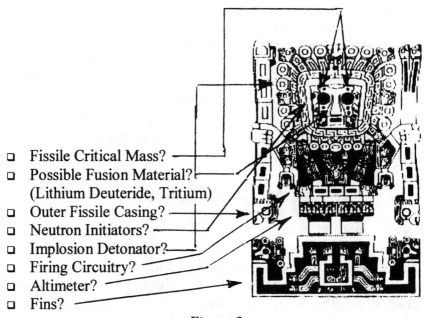

- Fissile Critical Mass?
- Possible Fusion Material?
  (Lithium Deuteride, Tritium)
- Outer Fissile Casing?
- Neutron Initiators?
- Implosion Detonator?
- Firing Circuitry?
- Altimeter?
- Fins?

Figure 2:
Viracocha: "A Hydrogen Bomb?"

The symmetry of the schematic along its central axis also suggests that this was intended to be the diagram of a three dimensional object that was round along the vertical axis. The analysis of the mural in terms of nuclear bomb engineering would suggest an aesthetic corroboration to Sitchin's thesis that the paleoancient Very High Civilization possessed nuclear weapons and used them in wars. It would also appear to corroborate the strong resemblance between passages in the Hindu epics and actual nuclear and thermonuclear detonations.[11]

2. Allusions of Circuitry in Occult Schematics

In his *Secret Teachings of All Ages*, occult writer and scholar Manley Hall presents various pentacles of the seven planets, their

---

[11] Cf. Chapter Three, "The Paleography of Paleophysics."

seals, and the characters of the planetary angels as they occur in the practice of ceremonial magic.

Such geometrical patterns recur in esoteric and occult writing, particularly in the systems of the Renaissance and afterward. It must not be assumed that neither the specific Hebrew characterization nor the zodiacal references are original to the tradition that created such geometries. However, certain similarities do emerge once the details of Hebrew and Jewish Qabalism are momentarily left out of view and the diagrams are examined simply for their geometric and schematic content.

- The angelic seals represent apparently meaningless "occult writing." But this writing also bears striking resemblance both to Ramsey graphs and to electronic circuitry schematics, and thus at one time may have had a specifically scientific and technical meaning long since forgotten.
- The planetary seals all have the names of four or three archangelic powers surrounding the geometric pattern of the planet.
- Every seal contains a character of the Hebrew alphabet on a surface of the geometric figures contained within the seal, and given the use of characters in the Hebrew to represent numbers, this might indicate something about the harmonics of hyperspace rather than the more mundane "mystical" explanation of various occult traditions.
- The resemblance of the geometries of the planetary seals with the spherically embedded Platonic solids is palpable.
- Only two seals contain the divine Name of God Himself, one in a system "in balance: and one in a system of "imbalance". That is to say, the primary intention of such "seals" appears to be simply to model various systems states.

The points, which interest us in this list of observations then, are the following:

- The constant recurrence of such complex geometric diagrams, usually in conjunction with writing or other symbolism, pictograph, or ideogram, and the zodiac; and,
- Why such geometric diagrams, from the Pythagoreans to Rosicrucianism and in other secret societies on down to our own day, would consider such common geometrical forms to be among their most closely guarded secrets? Even in the relatively sophisticated mathematical culture of the Renaissance this is the case, and so it appears reasonable to conclude that the mathematics and geometries themselves were not actually the secrets that were so zealously guarded, *but what that geometry and mathematics actually meant.* Since the occult meanings of such symbolism is fairly easy to come by and hence no great secret at all, perhaps the "secret" may be explained as the rudiments of an encoded science all but forgotten.

The answer to the second heading lies once again in the supposition of a paleoancient Very High Civilization, for which these geometries constituted a "key" not only to unlock their unified physics *but also to engineer it.* As such, such diagrams do not have anything to do with spiritual illumination or enlightenment in the usual esoteric sense, other than in the sense of understanding fundamental cosmological principles and forces, and their malign and benign uses. Indeed, the potential for malign and destructive uses of this knowledge constitutes perhaps the basic reason that such knowledge was concealed in the first place. I postulate that when the destruction of the society and infrastructure that made such engineering possible was immanent, that the paleoancient Very High Civilization began to encode this knowledge in the guise of esoteric metaphysical and religious teachings to be preserved by a select membership in an ongoing secret society tradition.

Turning then to a consideration of the diagrams themselves, certain analogies between planetary seals and angelic characters with modern electrical circuitry begin to evidence themselves. And this raises the question of exactly who introduced the contemporary ideogram and its meaning. Did this individual or individuals have any association with such societies?[12]

Planetary Seal or Angelic Character   *Electrical Schema*

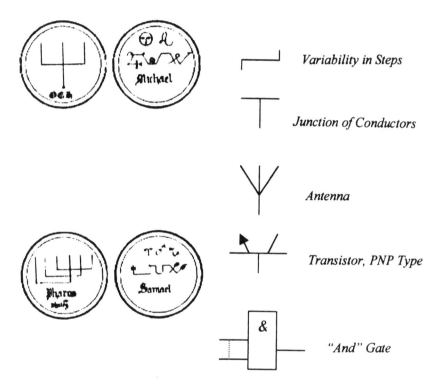

Variability in Steps

Junction of Conductors

Antenna

Transistor, PNP Type

"And" Gate

---

[12] In the case of Benjamin Franklin, whose electrical experiments are well known, one has a clear connection between a scientist and a secret society. The question then is, did Franklin introduce any ideogram drawn from some esoteric tradition, and if so, why did he choose a particular ideogram from that tradition to represent a certain limited and well-defined scientific and technical application?

Planetary Seal or Angelic Character   *Electrical Schema*

Alternating Current

Coherent Radiation

Ionizing Radiation

Thermal effect

Electromagnetic Effect

Resistor

Piezoelectric crystal: 2 electrodes

Semiconductor diode

C. Evidence of Secret Research into Paleophysics

Occult, conspiracy, and "alternative science" literature are full of wildly speculative rumors of secret research programs in places such as Area 51, or of secret "Marconi" or "Tesla" societies conducting their own research – free of the prying eyes of any government – in the wilderness of South America or elsewhere. And of course there are the standard and in some cases well-documented stories of secret Allied and Nazi research into occult and alternative sciences during World War Two.

> Atlantis is the missing piece in the puzzle of WWII, it is the beginning and ending of the war.
> If one makes the effort to unearth the myths, symbols and fantasies which comprise the mindsets of F.D.R., Hitler, Churchill and their associates, one will find that WWII was not "just" a war of good over absolute evil. It was a war of one occult "mind set" over another[13]

From the 19th and on into the 20th centuries until the outbreak of the war, American and German scholars poured over Near Eastern archaeological sites with a view to but one thing: proving their pedigree and connection with those ancient societies. It was the first outbreak of the "culture war", expressed in a kind of social imperialism as the ascendant Anglo-American culture and the ascending German culture each laid claim to a vaunted ancient pedigree. For the Anglo-Americans, the lineage ran from Egypt, to Greece, Rome, Britain and finally to America. For the Germans, the lineage ran from Atlantis, to Sumer, to India, and finally to Germany.[14] But proving that secret research into weaponized paleophysics was or is actually being conducted by contemporary governments or secret societies is no easy matter. Secret societies seem more bent on proving their pedigree from the Templars or some other murderous band of roving monks than on doing any

---

[13] William Henry, *One Foot in Atlantis: The Secret Occult History of world War II and Its Impact on New Age Politics* (Anchorage, Alaska: Earthpulse, 1998), p. 7.
[14] Ibid., p. 24.

serious scientific research into the possible meanings of their own texts.

There is, however, a certain body of evidence that suggests that research into paleophysics has been and is being conducted in secret, both by governments, and by secret societies. A pattern – small though persistent - does emerge of secret society interest in such esoteric physics, dating back to the Pythagoreans and Platonists (as we shall see) and winding its way through the Rosicrucians and other such societies down to our own day. In particular, it would appear that some Freemasons have made such research an ongoing component of their studies. With respect to ancient megaliths and monuments such as the Great Pyramid or Stonehenge,

> John Mitchell, in *City of Revelation,* comments that 'the traditions relating to these monuments are unanimous in claiming that they are relics of a former elemental science, founded upon principles of which we are now ignorant.' Among Masons the search for these lost principles has become an essential feature of their secret society. In November 1752, when George Washington became a Mason in Fredericksonburg, he heard the following words:
> "The proper business of a Mason is astronomical, chemical, *geological* and moral science, and more particularly that of the ancients, with all the mysteries and fables founded upon it.
> "Let us endeavor to *turn the stream*; to go from priest-craft to science, from mystery to knowledge, from allegory to real history."[15]

Of course, the capstone of the Great Pyramid, "said to have been made of pure crystal" and "believed to have attracted and transmitted cosmic rays,"[16] is a favorite symbol of Masonry, being

---

[15] Colin Wilson and Rand Flem-Ath, *The Atlantis Blueprint: unlocking the Ancient Mysteries of a Long-Lost Civilization* (New York: Delacorte, 2000), pp. xiii-xiv, emphasis added by Wilson and Flem-Ath. Wilson and Flem-Ath cite John Michell, (New York: Ballantine Books, 1972) *City of Revelation* , p. xiii.

[16] Henry, op. cit., p. 3. It should be noted that Henry does not cite a source for the allegation about transmitting cosmic rays. I mention it here, of course, because of its intrinsic interest to the weapon hypothesis.

included in the Great Seal of the United States, which Mason Franklin Delano Roosevelt ordered to be placed on the obverse of the one dollar bill in 1936.

More recently the Nazis are alleged not only to have investigated the use of zero point energy, which they called "Vril" energy, but also showed more than a passing interest in turning the Pythagorean understanding of harmonics and vibration into viable weapons systems.[17] And it is worth mentioning that the University of Chicago, founded with money from John D. Rockefeller, was of course heavily involved with the wartime Manhattan atomic bomb project and more recently in the development of alternative "Tesla" weapons technologies.[18]

The connection between big money and alternative science is, of course, one dating back to Morgan's funding both of Tesla and of Edison. But not so well known are the occult interests and connections of the three great "applied scientists" of the late nineteenth and early twentieth centuries. According to William Henry:

- Tesla was attempting to build a stargate.
- Thomas Edison was trying to construct a machine for communicating with the dead.
- Tesla's former assistant Marconi thought he had intercepted messages from Mars.[19]

But perhaps Tesla was investigating more. During his Colorado Springs experiments, Tesla's coil was a large square "barnlike structure one hundred feet long on each side. Its sides and roof sloped twenty-five feet into the sky. From the middle of the roof rose a wooden pyramid."[20] Why a Pyramid? The answer lies perhaps in Tesla's real purpose for these experiments. While they remain shrouded in as much mystery as the inventor himself, one

---

[17] Ibid., p. 14.

[18] Ibid., p. 22.

[19] Henry, op. cit., p. 31. Again, Henry makes no mention of his sources for these remarks, but the facts themselves are fairly well known.

[20] Ibid., p. 128.

Tesla biographer, John O'Neill, states that the goal of these awesome electrical displays was "merely to learn how the Planetary Grid would respond to his theory....Tesla wanted to know" whether or not, by applying his electrical "impulse" technology, "if the planetary grid was charged and if it could be set into electrical vibration."[21]

The occult tradition is also fairly consistent in the talismans of power, of which there seem to be four, dating back to the earliest texts and maintained with remarkable consistency over time and in several traditions:

- A magic cauldron (which may be the Holy Grail) called the Cup of Destiny;
- A magic sword called the Sword of Destiny;
- A magic spear or staff (which Hitler later claimed to have acquired) called the Spear of Destiny;
- A magic rock called the Stone of Destiny. (This Stone was believed to be the same stone that the biblical patriarch Jacob, the father of Dan, had used as a pillow when he had his vision of "Jacob's Ladder" which transported angels from heaven to earth.)[22]

We will encounter one of these talismans in the form of the "stones" or "magic crystals" that once were found inside the Grand Gallery of the Great Pyramid. What must be mentioned here is the amazing consistency of these four talismans not only across ancient cultural traditions but also through history. For example, the symbolism appears in the four minor suits of the typical Tarot deck as Cups, Swords, Wands, and Pentacles, or our modern playing card deck as Hearts, Spades, Clubs, and Diamonds.

But perhaps the most powerful indicator that there is indeed a secretive, and perhaps even secret society, influence on covert research into paleophysics is that Sitchin's own hypothesis of a "twelfth" planet of the solar system (12th if one counts the Sun and

---

[21] Ibid., p. 129.
[22] Ibid., pp. 28-29.

moon, as was the ancient custom), after it was first published, apparently became the subject of intense NASA scrutiny.

> In 1981, just five years after the publication of his first book, *The Twelfth Planet*, astronomers from NASA (among whose founding members were prominent former German scientists) and the US Marine Observatory in Washington D.C. were actively searching for this mysterious planet An (and presumably her alleged inhabitants).
>
> Simultaneously, geneticists backed by billions of dollars in public research money began decoding the Book of Life, our DNA, possibly in search of Enki's, the genetic artist's, "signature.."...
>
> In 1982, the space telescope IRAS (infrared astronomical station) saw what had not been seen for perhaps two thousand years. On December 30, 1983, the world press reported that the advanced space telescope:
>
> *"Discovered a celestial body in the direction of the constellation* **Orion**, *which possibly is as big as the gigantic Jupiter and perhaps so close to the earth that it could beglong to our solar system.... When IRAS researcheers saw the mysterious celestial body and calculated that it was possibly eighty billion kilometers away from earth, they speculated that it is moving towards earth."*
>
> In 1987, NASA confirmed what the ancient Sumerians knew: *"an eccentric tenth planet orbits the Sun."*[23]

Recently of course geneticists have confirmed another of Sitchin's hypotheses, namely, by demonstrating from human DNA material that we all stem from a common mother. But more intriguing is what is *not* mentioned by Henry: How did NASA *know* where to look, and *why* were they looking for it? The detection of a planetary body, especially one orbiting our own sun, is a long and time-consuming process. And that fact suggests that the NASA scientists were looking because they were prompted to look. Am I suggesting that someone in NASA was or is reading Sitchin? It seems a strong possibility.

Moreover, Face-on-Mars researcher Richard C. Hoagland has maintained in the past few years that there is a secret "NASA within NASA," a group governed by Masonic and other occult

---

[23] Henry, op. cit., p. 95, citing Michael Hesemann, *The Cosmic Connection* (Nath, UK: Gateway Books, 1996), p. 102, emphasis in the original.

influences and doctrines. Hoagland adduces support for this hypothesis from the strange occult symbolism found not only on the Mission Patches of the Gemini and Apollo astronauts' space suits, but from the timing of certain missions – particularly the Apollo missions – to certain Zodiacal alignments when viewed from the earth and the moon.

The notion of a "NASA within NASA" is not new to Hoagland, however. Indeed, with Hoagland, one has an individual who was once on the "inside" at NASA, and who came to this interpretation of events rather late in the game, and for very specific reasons. But the idea is not a new one. Ever since the assassination of President Kennedy, there has been in circulation a curious, and detailed, document known to most assassination researchers known as "the Torbitt Document," compiled by a Texas attorney with good political connections in Texas politics. Published in 1970, many of its astonishingly detailed allegations seemed outlandish at the time, but were subsequently confirmed. But one allegation remains curiously out of place in the document: the notion that there was a "secret" space program being conducted from Nevada under Werner von Braun and the Defense Advanced Research Projects Agency (DARPA). According to the document President Kennedy's attempt to interfere, curtail and acquire information from this project played a primary, if not the primary, factor in his assassination.

While none of these isolated cases of facts are significant enough proof in and of themselves, I do believe that taken together they constitute a significant enough body of evidence to suggest a *pattern* of ongoing secret society and/or governmental research into paleophysical matters. They constitute a *prima facie* case for the convening of a paleophysical "grand jury" to see whether or not the case should go to trial and be weighed on the evidence. One can easily guess at the governments that would be conducting such research: The United States, Russia, Japan, Germany, France, Britain, China, India, Canada, Italy - in short, all the world's technologically sophisticated "scientific power-houses." One may likewise guess at the secret societies that may be conducting similar research.

But getting those governments and institutions to "trial" will be difficult. And even if they were to allow it, one can only envision them "pleading the fifth" once they are in the witness stand.

III.
The Paleography of Paleophysics, Part One:
Thoth and Theories

"(A modern reader) does not think of the possibility that equally relevant knowledge might once have been expressed in everyday language. He never suspects such a possibility, although the visible accomplishments of ancient cultures – to mention only the pyramids or metallurgy – should be a cogent reason for concluding that serious and intelligent men were at work behind the stage, men who were bound to have used a technical language..."
Giorgio di Santillana.

A. New Interpretive Principles:
The Paradigm of Systems Entanglement Physics

If the Great Pyramid was a weapon of mass destruction it stands to reason that it employed principles of physics and engineering unknown to, or inadequately appreciated or ignored by contemporary physics. The method employed in this book is to extrapolate those physics principles based on what is already known in physics, either in currently accepted theory or past – and now rejected – ideas. Thus, our extrapolation will proceed along two tracks, one, the track of current theory and research, and the other, the track of ancient texts and the type of physics that might be implied by them.

This method is obviously not without its problems, for it is possible that the physics represented by the Pyramid was *so* advanced that a radical restructuring of our own theoretical models may be required to apprehend it. Likewise it is entirely possible that some new discovery may yet be found that would require the modification of part or all of the hypothetical reconstruction of that physics presented here.

However, a certain curious fact emerges, because most of the properties of the Pyramid only begin to take on scientific importance with the advance of theoretical physics since Newton, and exhibit an astounding correlation with the even more abstruse theories of the nineteenth and twentieth centuries. I have chosen to call this putative ancient "paleophysics" the physics of "Harmonic Systems Entanglement."[1]

The concepts and paradigms of the "physics of harmonic systems entanglement" form the basis for the examination in this chapter of the indicators, within certain ancient texts, that there was once an extremely sophisticated paleophysics based upon them. However, it must be understood that this chapter is *not* to be construed as the primary evidence on which that physics is postulated, nor as the ultimate foundation on which the pyramid weapons hypothesis rests.

Physicist Paul A. LaViolette states the case for this "principle of paleophysical interpretation" as follows:

> A thorough study of the various myths and lore carrying this science of creation indicates that it probably was not developed through loose analogy and speculative philosophy. Rather, its originators seem to have known precisely what they were doing. Available evidence shows the existence in ancient times of a quantitative science at least as sophisticated as our own, one that provided the basis for a relatively in-depth understanding of reaction system behavior and microphysical phenomena.[2]

The reason that these myths can only recently be understood as embodying a sophisticated physics is that "only recently has modern science formulated the theoretical foundations necessary ... with the emergence of the physics and chemistry subdiscipline known as *nonequilibrium thermodynamics....*"[3] Essentially this discipline, in LaViolette's hands, is used to explain how wave phenomena might account for fast-than-light transference of information between non-local systems. Here it suffices to remind the reader of the centrality of wave phenomena both to

---

[1] I use the term "paleophysics" throughout this book as the term identifying the weaponized application of the physics of the paleoancient Very High Civilization that built the Pyramid.

[2] Paul A. LaViolette, *Beyond the Big Bang: Ancient Myth and the Science of Creation*(Park Street Press: 1995), pp. 13-14.

[3] Ibid., p. 15.

LaViolette's model, and to what I have called the physics of harmonic systems entanglement.

With that in mind, we may now cite LaViolette on the necessary criteria for examining such paleographic evidence as may exist to discover the sophisticated paleophysics version of the cosmological model LaViolette and others believe to have been at work in those myths:

> In determining whether a given myth or lore shows a strong correlation with a modern creation physics paradigm,[4] the following three criteria should be applied. First, there should be some indication that the ancient myth or esoteric lore attempts to describe the origin of the universe or the process of physical creation. Second, the myth or lore must metaphorically express, through the described personality traits of its gods or characters, several specific concepts or principles corresponding to concepts or principles in the modern creation physics paradigm. Third, the myth or lore should present these metaphorical traits or actions in a sequential order that corresponds to the logical order of events inherent in the creation physics paradigm.[5]

The only problematical requirement that would seem to be in evidence here is the third, for if these myths and lore are indeed the products of any given "legacy" civilization trying to understand the more advanced physics of its more sophisticated predecessor, then it is likely that the texts themselves may not be exact in their preservation of this sequential order, a fact compounded if one takes into consideration the normal and natural corruption of textual transmission.[6] LaViolette's third requirement would

---

[4] By "creation physics paradigm" LaViolette means precisely the nonequilibirum thermodynamics. More broadly, however, he also means to imply some of the salient implications suggested by other developments with physics -- such as chaos theory and plasma cosmology, which are reviewed in chapter three – and some of the physics implications suggested by systems research in biology.

[5] LaViolette, op. cit., p. 16.

[6] This is my first, and last, bow to the standard assumptions of textual transmission and reconstruction. That they are entirely western European based, and contradicted by any number of extant oral traditions preserved for generations in many "primitive" cultures, is well known. I do not share those assumptions nor am I in sympathy with them.

therefore appear to necessitate some modification along the following lines: *Third, the myth or lore should (a) either present these metaphorical traits or actions in a sequential order that corresponds to the logical order of events inherent in the creation physics paradigm; or (b) in such a fashion as to allow the algorithm of that paradigm to be significantly reconstructed from the extant texts; or (c) to allow the theoretical assumptions of that paradigm to be reconstructed.*

B. Ancient Egyptian Pyramid Texts and Sirius

As a tantalizing way to afford entry into an examination of the paleography of paleophysics, many have pointed out the connection between the Great Pyramid's celestial alignment with and duplication of the constellation Orion (and in particular the star Sirius) in the ground plan of Giza itself. Those correlations are also reproduced in ancient Egypt's pharaonic death-resurrection religion.[7] This correspondence of celestial alignment and terrestrial religion, plus the corroborative evidence of the ancient Egyptian "Pyramid Texts" themselves, is some of the most persuasive evidence marshaled in favor of the various versions of the Pyramid functioning as some sort of observatory, the "Observatory Hypothesis." Yet I am bold to suggest that perhaps two of the most obvious facts about that correspondence have been overlooked (not to mention the suggestive paleographic evidence assembled by Sitchin that the Great Pyramid was indeed a weapon):

(1) the association of Giza, celestially, with a Zodiacal constellation whose primary metaphorical and religious significance was *death*; and ,

(2) the embodiment of that association in the "as above, so below" principle of systems entanglement in so much evidence at Giza.

---

[7] Cf. Robert Bauval and Graham Hancock, *The Orion Mystery.*

41

Considered as a whole, the paleographic evidence as touching directly upon the pyramid complex at Giza, as well as the religious associations of Egyptian mythology, point persuasively to the Weapon Hypothesis as to no other. It must be constantly remembered that the Egyptian myths themselves are the creations of the legacy civilization, a distant "harmonic" of what was once a loud, and purely scientific, "fundamental."[8]   This conclusion becomes even stronger when the broader paleographic evidence for the existence in very ancient times of some version of systems entanglement physics is considered. This conclusion would seem to be finally persuasive when the parameters of that physics are considered by a direct analysis of the Pyramid and the Giza complex.

C. *Zechariah Sitchin and the Great Pyramid as a Weapon*

Zechariah Sitchin is a researcher who, like Immanuel Velichkovsky, hovers on the fringes of science in some pseudo-scientific territory, proposing, on the basis of a detailed linguistic analysis of ancient texts, a rather far reaching   hypothetical reconstruction of the history of the ancient High Precursor Civilization, including, among other things, its extra-terrestrial origins! This does not, however, invalidate the individual and oftentimes unique insights that go to comprise the individual components of his total hypothesis, one of the most compelling of which is his examination of texts that suggest that the Great Pyramid's primary function was as some sort of weapons platform of mass destruction. This would imply two things, both of which,

---

[8] This of course requires the non-orthodox Very Early Dating for the Great Pyramid and some other structures of the complex (i.e., that *none* of the main structures of Giza were built by any of the Pharaohs), but much, much earlier by the "precursor" civilization so often alluded to by other ancient civilizations such as Sumer, the Indus valley civilization, and, not surprisingly, Egypt itself. "Orthodox Egyptology" consistently ignores the fact that the ancient Egyptians viewed themselves as a legacy, and ascribed a state of extreme scientific and technological advancement to their ancestors.

as will be shown, corroborate Sitchin's "weapons texts", for if the Pyramid was a weapon, then:

(1) it was, by definition, some sort of *machine*, a fact which can be verified by careful consideration of its (remaining) components and construction, a task best exemplified in the work of Christopher Dunn; and,

(2) *other* texts not examined by Sitchin should indicate a similar purpose or function of the Giza complex. Failing that, they should corroborate it is a general fashion by indicating in some fashion that there once existed sophisticated weapons of mass destruction in ancient times.

Thus, the methodology being adopted here does *not* require commitment to Sitchin's overall model. Rather, taking the texts cited by Sitchin as the indicator of the Weapon Hypothesis, we seek to corroborate that hypothesis by discovering:

(1) whether or not advanced physics and/or engineering principles are alluded to in other ancient texts;

(2) whether or not those principles are consistent across several cultural contexts;[9] and,

---

[9] This requires some additional comment. The legacy civilizations of Indus, Sumer, and Central and South America, not to mention some North American native Indian traditions, African tribal tradition, and Polynesian and Oriental traditions, all bear witness to some more advanced precursor Ancient High Civilization which preceded them. More peculiarly, these traditions overlap to a degree of detail that suggests that this paleoancient Very High Civilization was global in the extent of its technological embrace and impact, much like our own. This would consequently imply that the paleophysics itself was embodied in similar fashion and detail, hence the examination in this chapter of Egyptian, Sumerian, Greek and Sanskrit texts. In the case of the examination of the Egyptian and Sumerian texts, I am reliant upon the work of Hancock and Sitchin. In the case of Plato, I rely upon LaViolette's analysis, but extend its implications considerably to suggest a sweeping reinterpretation of the whole Platonic corpus. In the case of the *Hermetica* of Hermes Trismegistus, the

(3) whether or not those principles are embodied at the Giza complex in a manner that would suggest their weaponization.

Sitchin presents credible textual evidence that the Great Pyramid of Giza was the primary component of some form of "paleoancient", though certainly not primitive, weapons system of mass destruction in his work *The Wars of Gods and Men.*[10] This section is a precis of that textual evidence and a brief analysis of the type of weapon that his analysis of the texts seems to imply.

Sitchin's texts relate to the final events of a paleo-global war, apparently fought, according to him, with nuclear and other more terrifying weapons of mass destruction, a war similar in this respect to similar wars recounted in the ancient Hindu epics, the *Ramayana* and the *Mahabharatra*. Sitchin call this war "the Second Pyramid War". The focal point of this war is for control of the Great Pyramid, which is the ultimate weapon. This in itself suggests that its destructive power was far in excess of that imaginable with nuclear weapons, since such weapons were, according to Sitchin, deployed in that war and *used in a subsequent war* after the Pyramid's destruction.

The outlines of that struggle go something like this. The gods of Mesopotamia, eventually victorious over Marduk, who was besieged within the Pyramid itself, dispatch a team to enter the structure, inventory its contents, and designate which components should be destroyed and which should be waved for utilization in other devices.

Thus, if the Pyramid was a weapon, then (adding to our previous list of verifiable tests):

---

examination is entirely my own, though suggested by Bauvall and Hancock's work.

[10] Zechariah Sitchin, *The Wars of Gods and Men,* Book III of *The Earth Chronicles* (New York, Avon Books: 1985), pp. 163-174.

(1) it was, by definition, some sort of *machine*, a fact which can be verified by careful consideration of its (remaining) components and construction; and,

(2) *other* texts not examined by Sitchin should indicate a similar purpose or function of the Giza complex; and,

(3) some of its components are missing from the structure, a fact which should be in evidence by a careful consideration of the contemporary shell now remaining at Giza.

This list constitutes the first series of Sitchin's "pyramid hypotheses," and it must be carefully considered before proceeding.

First, the texts imply that the structure was conceived and built as a machine, in this case, a weapon of extraordinary power. As such, the verification of this hypothesis will lie in part in an examination of the *form* and *of the materials used* in its construction. Given the number of mathematical and harmonic relationships *alone* that are embodied within it, I postulate that the simplest structure that *could* be built embodying all these relationships in precisely the manner that they were embodied was precisely the form of a pyramid *and no other.*[11]

Second, Sitchin's texts indicate that the structure was subsequently entered after its completion for the purpose of an inventory and destruction of some of its contents and removal of other components. The verification of this hypothesis will lie in any trace evidence that the pyramid was entered prior to its "modern" forced entry by the Moslem caliph in the ninth century. As will subsequently be seen, Sitchin's texts moreover imply the manner of this entry, and thereby the manner of its verification.

Third, the very suggestion of missing components means the modern structure is but a shell of its true, former, self. This

---

[11] That I cannot, obviously, test this postulate goes without saying, for it would require the modeling and computational power of a very large mainframe computer and considerable care and expertise in establishing the parameters of the program.

indicates that *some of its functions cannot properly be understood without an exact knowledge of its missing components and their functions.* Such components as may have been deposited initially within the structure must be speculatively reconstructed on the basis of the existing structure and the physical functions of its various components and nested mathematical, harmonic relationships. Sitchin's texts do provide an important clue as to what those missing components may have been. His texts indicate that the Great Pyramid's primary function was that of a weapons platform. Moreover, they indicate that it was a weapon of such destructive power as it exceeded the power of nuclear weapons. They indicate that its destructive power was so extraordinary that the "victors" in the "Second Pyramid War" ordered its permanent incapacitation, then this means that the shell that remains at Giza is the *secondary* structure. The *primary* components are missing.

Consequently, any theory that attempts to reconstruct its function solely on the basis of "reverse engineering" its remaining shell without consulting those texts, and therefore without due consideration of what those missing components might have been, is an inadequate theory. In this respect, Baauvall, Hancock, and others are correct. The ancient religious texts *are* crucial to a proper understanding of the Pyramid's ultimate purpose and function.

Sitchin's texts and his analysis are now reproduced here in the order of his presentation.

(1)
General Weapons Properties of the Great Pyramid:

We learn more of the last phases of this Pyramid War from yet another text, first pieced together by George A Barton (*Miscellaneous Babylonian Texts*) from fragments of an inscribed clay cylinder found in the ruins of Enlil's temple in Nippur.

As Nergal joined the defenders of the Great Pyramid ("the Formidible House which is Raised Up Like a Heap"), he strengthened its defenses through various *ray emitting crystals* (mineral "stones") *positioned within the Pyramid.*

> "The Water-stone, the Apex-stone,
> the...-stone, the...
> ...The Lord Nergal
> increased its strength.
> The door for protection he... to Heaven its Eye he raised,
> Dug deep that which gives life...
> ...in the house he fed them food."[12]

It should be noted that the interpretation of these magic "stones" as "ray-emitting crystals" is Sitchin's own. We shall encounter a different understanding of what these stones may have been with Christopher Dunn's version of the Machine Hypothesis in chapter five.

(2)
Its Apparently Radioactive or Strong Electromagnetic Field Properties:

> Ninurta was at first astounded by her decision to "enter alone the enemyland"; but since her mind was made up, he provided her with "clothes which should make her unafraid" (*of the raditation left by the beams?*). As she neared the Pyramid, she addressed Enki: "She shouts to him... she beseeches him." The exchanges are lost by breaks in the tablet; but Enki agreed to surrender the Pyramid to her:
>
>> "The House that is Like a Heap,
>> That which I have as a Pile Raised Up –
>> Its mistress you may be."
>
> There was, however, a condition: the surrender was subject to a final resolution of the conflict until "the destiny-determining time" shall come. Promising to relay Enki's conditions, Ninhursag went to address Enlil.[13]

---

[12] Sitchin, op. cit., pp. 163-164, emphasis added.
[13] Ibid., p. 165, emphasis added.

Again, the conclusion that Ninhursag wore radioactive protective clothing is a conclusion that Sitchin himself makes, based upon his examination of this and numerous other texts. As will be discovered in examining Dunn's Machine Hypothesis, it is a conclusion *warranted* by some of the likely purposes of the structure itself.

<u>(3)</u>
The Motivation of the War:

Nowadays the visitor to the Great Pyramid finds its passages and chambers bare and empty, its complex inner construction apparently purposeless, its niches and nooks meaningless.

It has been so ever since the first men had entered the Pyramid. But it was not se when Ninurta entered it – circa 8670 B.C. according to our calculations. "Unto the radiant place," yielded by its defenders, Ninurta had entered, the Sumerian text relates. And what he had done after he entered changed not only the Great pyramid from within and without but also the course of human affairs.

When, for the first time ever, Ninurta went into the "House which is Like a Mountain," he must have wondered what he would find inside. Conceived by Enki/Ptah, planned by Ra/Marduk, built by Geb, equipped by Thoth, defended by Nergal, what mysteries of space guidance, what secrets of impregnable defense did it hold?"[14]

---

[14] Ibid. It should be noted that one component of Sitchin's wider "extra-terrestrial Model" now intrudes. For Sitchin, the Pyramid also functioned as some sort of communications device guiding "ancient astronauts" to a spaceport in Sumeria. He wavers back and forth between these two functions, though his *texts* clearly indicate its primary function as a weapon. Sitchin offers no explanation of how it might possibly have done both. Both are possible, particularly if it was a certain *type* of weapon embodying a certain *type* of physics.

(4)
The Focus of Nergal's Interest

A striaght descending passage led to the lower service chambers where Ninurta could see a shaft dug by defenders in search for subterranean water.[15] But his interest focused on the upper passages and chambers; there, *the magical "stones" were arrayed – minerals and crystals; some earthly, some heavenly, some the likes of which he had never seen. From them there were emitted the beamed pulsations for the guidance of the astronauts and the radiations for the defense of the structure.*

Escorted by the Chief mineralmaster, Ninurta inspected the array of "stones" and instruments. As he stopped by each of them, he determined its destiny – to be smashed up and destroyed, to be taken away for display, or to be intalled as instruments elsewhere. We know of these "destinies", and of the order in which Ninurta stopped by the stones, from the text inscribed on tablets 10-13 of the epic poem *Lugale-e*. It is by following and correctly interpreting this text that the mystery of the purpose and function of many features of the pyramid's inner structure can be finally understood.[16]

Having stated this, Sitchin then fails to list the names of the very stones he cites as being so crucial to an understanding of its purpose and function!

---

[15] This requires some comment. Sitchin is doubtless referring to the so-called "Well Shaft" leading from the Grand Gallery. This notion strains credibility, because the inside chambers of the Pyramid, during its operation in the relatively "benign" function Dunn ascribes to it, would have been uninhabitable, nor in its Weapon mode would it have been necessary for "defenders" to be located physically within it, any more than humans live *inside* the nuclear reactors. However, there is some merit to Sitchin's suggestion in that *if* the Weapon Hypothesis is true, the likelihood does exist of yet undiscovered (or undisclosed) chambers and passages far beneath the Giza complex. This idea is strengthened by the allegations of some that not only do such passages exist.

[16] Sitchin, op. cit., pp. 167-168, emphasis added.

Going up the ascending passage, Ninurta reached its junction with the imposing Grand Gallery and a horizontal passage. Ninurta followed the horizontal passage first, reaching a large chamber with a corbelled roof. Called the "Vulva" in the Ninharsag poem, this chamber's axis lay exactly on the east-west (axis) of the pyramid. Its *emissions ("an outpouring like a lion whom no one dares attack") came from a stone fitted into a niche that was hollowed out in the east wall. It was the SHAM ("Destiny") stone. Emitting a red radiance which Ninurta "saw in the darkness," it was the pulsating heart of the pyramid. But it was anathema to Ninurta, for during the battle, when he was aloft, this stone's "strong power" was used "to grab to pull me, with a tracking which kills to seize me."* He ordered it "pulled out... to be taken apart...and to obliteration be destroyed."[17]

If Sitchin is correct that the "stones" were some sort of crystals, then it is likely that they were *artificial* crystals since:

(1) they fit precisely into certain spaces made for them;
(2) and apparently could be "taken apart" before their final destruction, presumably so they could be physically removed from the structure, taken elsewhere, and their destruction completed under supervision.

We will return to this point in chapter six.

The other alternative interpretation is that "stones" represents a more corrupted attempt by the legacy civilization to describe some machinery of a complicated nature that perhaps involved "stones" or "crystals" in some fashion. As will be seen, Dunn inadvertently corroborates Sitchin's "crystals" reading of "stones" in his own version of the Machine Hypothesis. In any case, it is apparent that Sitchin's texts indicate that the missing components in the interior

---

[17] Ibid., p. 168, emphasis added.

of the Pyramid involved some sort of function necessitating the use of crystals and possibly of an optical cavity. Both strongly suggest its function involved harmonics to a high degree.

(6)
The "Grand Gallery" and Ninurta's Ascent:

> Returning to the junction of the passage, Ninurta looked around him in the Grand Gallery.... Compared to the low and narrow passages, it rose high (some twenty-eight feet).... Whereas in the narrow passages only "a dim green light glowed,"[18] the Gallery glittered *in multi-colored lights—"its vault is like a rainbow, the darkness ends there." The many-hued glows were emitted by twenty-seven pairs of diverse crystal stones that were evenly spaced along the whole length of each side of the Gallery. These glowing stones were placed in cavities that were precisely cut into the ramps that ran the length of both sides of the Gallery on both sides of its floor. Firmly held into place by an elaborate niche in the wall, each crystal stone emitted a different radiance, giving the place its rainbow effect.* For the moment Ninurta passed by them on his way up; His priority was the uppermost chamber and its pulsating stone.[19]

As will be seen, Dunn comes to a rather different conclusion as to what fit in the niches cut into the ramps, based solely on an engineer's examination of the pyramid. He believes that these slots or niches held banks of Helmholtz resonators. As will be argued in chapter six, in order to function effectively as a Weapon, the Grand Gallery would have had to incorporate some version both of Dunn's and Sitchin's apparatuses: artificial crystalline Helmholtz resonators that were both optically *and* acoustically resonant.

---

[18] Ionized atmosphere gives off such a glow. And as we shall see, the Pyramid was filled with ionized hydrogen plasma.

[19] Ibid., p. 168, emphasis added.

(7)

The "King's Chamber and its "Coffer: The "Net" of Celestial Coupling:

He was now in the Pyramid's *most restricted ("sacred") chamber, from which the guiding "net" radar?) was "spread out" to "survey heaven and earth." The delicate mechanism was housed in a hollowed out stone chest;[20] placed precisely on the north-south axis of the Pyramid, it responded to vibrations with bell-like resonance. The heart of the guidance unit was the GUG("direction determing") stone; its emissions, amplified by five follow compartments[21] constructed above the chamber, were beamed out and up through two sloping channels leading to the north and south faces of the Pyramid. Ninurta ordered this stone destroyed: "Then by the fate-determining Ninurta, on that day was the GUG stone from its hollow taken out and smashed."[22]*

(8)

Commentary: The Parabolic Reflecting Faces of the Great Pyramid and a Common Mistake:

At this juncture, it is necessary to interrupt the presentation of Sitchin's textual data to focus on what is a common misperception, and that is that the "air shafts" leading from the "King's Chamber" (and by implication, the similar shafts leading from the "Queen's Chamber") served the function – for those who accept *any* version of the Machine Hypothesis (Weapon, Observatory, Communication or otherwise) – of emitting beams *from* the Pyramid outward. There is a property of the Great Pyramid, *unique to it alone of all the Pyramidal monoliths on the earth,* and that is that each of its

---

[20] I.e., the "coffer".

[21] Again, as will be shown, this is only partially correct, as the amplification effect was actually achieved by the huge granite stones that comprise the roofs of those chambers, and not by the hollow chambers themselves.

[22] Sitchin, op. cit., p. 169, emphasis added.

four faces is indented slightly at the center; *each face, in effect, constitutes a parabolic reflector much like a modern satellite dish.* As will be shown, only Christopher Dunn has correctly perceived that at least *one* of these shafts must be for the purpose of signal or energy *input.* This fact has a profound implication for the Weapon Hypothesis and to the interpretation of the rest of the Giza complex:

(1) parabolic reflectors are designed primary to *receive, collect, and amplify signal input, not to send to transmit signals.* This implies that *one primary function of the Great Pyramid is as a collector and amplifier.* The question is, what kind of signals are being collected and amplified, and for what purpose?[23]

(2) Satellite dishes and radio telescopes have an amplifier fixed at the focal center of the reflector which collect and amplify the reflected signals. This has further implications:

(a) part of the missing structural components of the pyramid either at one time resided *outside* the structure at some focal point in front of one, (or most likely, all) of its faces in a manner to a signal collector on a satellite dish. This would imply that some trace of such external structures as once may have existed should be evident, or may once have been evident, in the complex; *or,*

(b) those missing structural components still exist in the other structures of the complex, but their function has not yet been adequately perceived; *or,*

(c) the four faces of the Pyramid reflect an advanced engineering principle *making the Pyramid itself that "collector and amplifier" – the view that will be*

---

[23] Needless to say, the "extra-terrestrial" hypothesis is totally unnecessary here, since its could have simply been collecting and amplifying the signal of the background radiation of space itself, much as would a large modern radio telescope. This is precisely what I argue in chapter six.

defended subsequently – and further suggesting that none of the "air shafts" served as the primary output of whatever signal the Pyramid emitted, such output being the function, perhaps, of the missing apex "stone."[24]

With this in mind, we may resume our survey of Sitchin.

(9)
The "Grand Gallery" and Ninurta's Descent:

Now came the turn of the mineral stones and crystals positioned around the ramps in the Grand Gallery. As he walked down Ninurta stopped by each one to declare its fate. Were it not for breaks in the clay tablets on which the text was written we would have had the names of all twenty-seven of them; as it is only twenty-two names are legible. *Several of them Ninurta ordered to be crushed or pulverized;* others, which could be used…were ordered given to Shamash; and the rest were carried off to Mesopotamia, to be displayed in Ninurta's temple in Nippur, and elsewhere, as constant evidence of the great victory….

All this, Ninurta announced, he was doing not only for his sake, but for future generations, too: *"Let the fear ot thee" – the Great Pyramid – "be removed from my descendents; let their peace be ordained."[25]*

---

[24] A common question that is asked me whenever I have talked publicly about the Pyramid as a weapon is where its "signal" was emitted or "How was it aimed?" This involves a massive misperception of the type of weaponry and physics involved, for it assumes that the Pyramid was some sort of "directed energy" weapon, like a particle beam or a laser, a view strengthened by Dunn's very cogent and persuasive argument that the "King's Chamber" employed a maser in its construction. This was *not*, however, the type of weaponry or physics ultimately involved, since the directed energy component was only utilized to access the far more powerful potentials latent in the energy potential of the geometry of space itself. This means that its primary energy output came in the form of *non*-linear directed energy.

[25] Sitchin, op. cit, p. 171, emphasis added.

(10)
The Capstone:

Finally there was the Apex Stone of the Pyramid, the UL ("High as the Sky") stone: "Let the mother's offspring see it no more," he ordered. And, as the stone was sent crashing down, "Let everyone distance himself," he shouted. The "stones," which were anathema to Ninurta, were no more.[26]

This is an important piece of information, for it indicates that whatever else the missing Apex "stone" may have been, it was crucial to the function of the Pyramid *as a Weapon*. This constitutes, in my opinion, the strongest reason to speculate that it involved non-linear directed energy, and that its main "output" was through, and possibly in part directed by, the missing Apex.[27]

(11)
The Victory Commemoration Seal:

The Second Pyramid War was over; but its ferocity and feats, and Ninurta's final victory at the Pyramids of Giza, were remembered long thereafter in epic and song – and in a remarkable drawing on a cylinder seal, showing Ninurta's Divine Bird within a victory wreath, soaring in triumph above the two Great Pyramids.[28] (Figure One)

Figure One: The Victory Seal

---

[26] Ibid.
[27] Cf. chapter six.
[28] Sitchin, op. cit., pp. 171-172.

(12)
The Destruction of "the Great Weapon":

After Ninhursag had finished her oracle of peace, Enlil was the first one to speak. *"Removed is the Affliction from the face of the earth,"* Enlil declared to Enki; *"the Great Weapon is lifted up."*[29]

D. *LaViolette and Plato: The "Reaction-Diffusion Wave" of "Atlantis"*

Physicist Paul A. LaViolette presents yet another convincing case for the paleographic encryption of an ancient paleophysics, in this case, in the *Critias* and *Timaeus* dialogues of Plato. In this instance, however, we will present a case that goes beyond LaViolette's analysis, and indeed that of most conventional academic philosophy, by arguing that the "Atlantis" sections of these two dialogues may constitute a central core of the Platonic theory of universals or ideal forms, and not just an interesting tangential story.[30]

That case, however, most be understood against the background of LaViolette's "systems kinetics" (examined in the next chapter) as also against the backdrop of his wider criteria for examining paleographic evidence for this ancient paleophysics. But we may outline his presentation as follows. As recounted in those two dialogues, Atlantis flourished until it came to its catastrophic end ca. 11,600 years ago. "The tale, related by Socrates' elder student Critias, was originally told to Solon, ruler

---

[29] Ibid., p. 174, emphasis added.

[30] The most breathtaking example of an encoded paleophysics in Plato is the extraordinarily erudite academic study by Ernest G. McClain entitled *The Pythagorean Plato: Prelude to the Song Itself* (York Beach, Maine: Nicolas-Hays, Inc., 1984). I cannot recommend this highly detailed study enough. Suffice it to say that McClain presents all but overwhelming evidence that something like our modern western musical system of "tempering" the musical scale was known by *someone* in very ancient times.

of Athens, by priests he met in the Egyptian city of Sais."[31]   At this
point it is necessary to cite LaViolette himself at length:

> Any attempt to decipher the symbolic meaning of the Atlantis myth
> must explain why Plato's dialogues divide the myth into two portions.
> The first part is in the dialogue *Critias*, which relates the story of how
> Atlantis was created and gives a detailed description of the physical
> layout and commercial activities of this metropolis. The second part,
> given in the *Timaeus* dialogue, describes how the people of Atlantis
> waged war on the antediluvian Hellene civilization and were finally
> destroyed in a worldwide flood. It is here that reference is made to the
> "sinking" of Atlantis.
>
> Of the two narratives, the story in the *Critias* proves to be of
> particular interest from the perspective of systems science and the
> ancient ether physics.   Like the creation myths... the legend of
> Atlantis's creation encodes a sophisticated open-system ether physics
> that describes how the first particle of matter came into being from the
> etheric sea. *It even encodes a diagram showing how the primordial
> particle's energy field intensities vary as a function of distance from
> the particle's center. In some ways, the Atlantean creation myth
> presents one of the most sophisticated and graphic portrayals of this
> ancient creation science.*[32]

More will be stated momentarily about the connection of this
encoded ether paleophysics and its connection to the Platonic
"Allegory of the Cave". For the moment, attention must be focused
on LaViolette's summary of the Atlantean material of the *Critias*
and its meaning for physics:

> When (the gods) portioned out the earth, Poseidon received for his lot
> an island in the middle of the ocean. This was Atlantis. It had a
> diameter of roughly one hundred stadia(one hundred furlongs, or about
> twenty kilometers) and consisted for the most part of a very fertile plain
> in the center of which was a mountain of modest hieght. In this
> mountain lived two mortals, Evenor and his wife, Leucippe, both
> offspring of Poseidon. These two mortals had a daughter named Clito.
>
> One day Clito's father and mother died, leaving her alone when she
> had barely reached womanhood. Poseidon, desiring this maiden, made
> love with her in her mountain abode. Thereupon he fortified the

---

[31] LaViolette, op. cit., p. 220.
[32] Ibid., p. 221, emphasis added.

surrounding territory by re-forming the ground so that alternating rings of sea and land enclosed the central hill where she dwelt. There were two rings of earth separated by three rings of sea, all concentric with one another so as to form a bull's-eye pattern.[33]

LaViolette then cites the *Critias* directly for the exact dimensions of this pattern:

> The breadth of the largest ring of water... was three stadia and a half, and that of the contiguous ring of land the same. Of the second pair, the ring of water had a breadth of two stadia and that of land was once more equal in breadth to the water outside it; the one which immediately surrounded the central iselt was in breadth one stadium; the islet on which the palace stood had a diameter of five stadia.[34]

He the produces the following diagram of "Atlantis":

Viewed as an encoded principle of an ancient paleophysics, the meaning is plain, *but only to a scientifically and technologically sophisticated society:* "In the context of the ancient ether physics, Atlantis's land contour wave pattern charts the electric field potential that would compose the stationary wave profile of a proton."[35]

---

[33] Ibid.

[34] Plato, *Critias* 115e-116a, in *Collected Dialogues of Plato*, ed. E. Hamilton and H. Cairns, trans. A.E. Taylor (Princeton, N.J.: Princeton University Press, 1961), cited in LaViolette, op. cit., p. 223.

[35] Ibid.

More importantly, viewed from the Brusselator systems analysis model LaViolette propounds (and reviewed in the next chapter),

> Poseidon(water) and Clito (land) symbolize the two interacting X and Y ether variables that are cross-coupled into a self-closing transforming loop. Poseidon's breaking of the land contour symmetry to form alternating concentric rings of land and water (elevated land versus depressed land) illustrates how the X and Y ethers depart from their initial steady-state concentrations to form a wavelike concentration pattern configured as a series of concentric shells.[36]

A somewhat radical interpretation of the whole platonic system of universal forms emerges from this, since paleophysics and the Atlantean allegory are taken as the central core elements of the platonic system.

This radical reinterpretation may best be approached by exhibiting how the platonic "Allegory of the Cave" and its related doctrine of the soul and its faculties and objects of intellection are entirely recast.

The famous Roman Catholic scholar of philosophy Frederick Copleston reproduces the following tabular summary of Plato's doctrine of the soul, its faculties, and its objects of knowledge.[37]

Mental State		Objects Known	
Scientific	Intellection	Sources, Principles	Invisibles
Understanding	(ν νοησις)	(αι αρχαι)	(τα αορατα)
(η επιστημη)			
Knowledge	Discernment	Mathematicals	Intelligibles
(η γνωσις)	(η διανοια)	(τα μαθηματικα)	(τα νοητα)
	Belief, Faith	Living Things	Visibles
	(η πιστις)	(ζωα, κ.τ.λ.)	(τα ορατα)
Opinion-glory			
(η δοξα)			
	Shadows, Images	Icons	Perceivables
	(η εικασια)	(εικονες)	(δοξαστα)

---

[36] Ibid., p. 222.

[37] Frederick Copleston, S.J., *A History of Philosophy*, Vol. I, Part I, *Greece and Rome* (Garden City, N.Y.: Image Doubleday, 1962), p. 176.

The compelling thing suggested by this table are the scientific and physics principles embodied in the top two layers, for as the diagram indicates, the "Platonic turn" (περιαγωγη) consists in turning from the lower soul and its objects, up through "Living Things" through "Mathematicals" and finally to arrive at the Sources or Principles underlying things. That is to say, *a truly scientific understanding of physical mechanics (the underlying "universal principles" of things) requires a methodology of three basic steps, (1) the turn from mere icons to (2) a perception of "living things" (open, entangled systems), to (3) "mathematicals", the mathematical modeling of the principles underlying the invisibles (the quantum and sub-quantum) world.* What has usually been understood as an allegorical journey of the soul toward enlightenment is really the description of a particular kind of enlightenment based on a cursory outline of the scientific method in physics. And a *very* sophisticated physics it is, as we shall discover.

The implication is that the Platonic system is exactly what the Neoplatonists actually insisted it was, a deliberately encoded system containing hidden or occult truths. But those truths were not what the Neoplatonists, or for that matter Aristotle, took them to be. They were not truths of a primarily religious or metaphysical nature, but encoded physics. That being said, the implication is enormous, for the Aristotelian and Neoplatonic understanding of Plato – and most academic philosophy since then – may be the most egregious case of misunderstanding in history. If this view of Plato – somewhat radical though it is – is correct, then the underlying forms that Plato maintained underlie "this chair" and "that chair" was not a "superchair" but a *topological principle, i.e., a physico-mathematical law of form.*[38] This type of hermeneutic

---

[38] This case is aided, rather than impeded, by the claim of the Neoplatonist Proclus that Plato himself was an initiate into the "Egyptian mysteries", for this would imply a certain deliberateness in Plato's habit, not only of scattering certain salient features of his system in different works but of employing a deliberately allegorical rhetorical style.

based on open entangled systems will be the basis of our examination of the *Hermetica* of Hermes Trismegistus in the next section.

These considerations allow us to see more clearly what physics principles the religious and philosophical texts may have been trying to communicate or preserve. Plato's Allegory of the Cave, not to mention the whole ancient religious preoccupation to "free" the soul from the constraints of the material body, may be yet another profound misunderstanding on the part of the legacy civilizations of what was really being symbolized. If one understands the Allegory of the Cave as being another version of the Platonic turn from the lower soul, through "living things" or open entangled systems, to the mathematical principles underlying invisible things and giving rise to their forms, then Plato may indeed have been some kind of initiate, for this is an accuracy of knowledge rare for the day, best explained, perhaps, by someone "on the inside". On this reading of Plato, the material, visible world, down to the smallest quantum interactions, is itself the result of more fundamental reactions taking place in an invisible active substrate that may be precisely geometrically modeled, provided that the revolutionary "turn" or change in mental outlook takes, as its first step, the view of all things as "living", as open entangled systems.

E. The "Hermetica" and Hancock

Authors Graham Hancock and Robert Bauvall are two other investigators who maintain that ancient texts, particularly those regarding Egypt and/or the Giza Pyramid, should be investigated to ascertain the remnants of a sophisticated, ancient, and antediluvian, "paleophysics." As an entrance into a discussion of these ancient texts, known as the *Hermetica* of Hermes Trismegistus, the following questions are posed: "But why should the ancients have sought to create a similacrum of the skies on the ground at Giza? Or, to put the question another way, why should they have sought to bring down to earth an image of the heavens?"[39] Their book, *The Message of the Sphinx*, largely written to answer these questions, is yet another variation -- and a well-argued one – on the Giza complex as-ancient-observatory, with a modestly stated extra-terrestrial theme. Yet the texts they immediately adduce suggest a more sinister purpose behind the construction of the Giza complex:

> There exists an ancient body of writings, compiled in Greek in the Egyptian city of Alexandria in the early centuries of the Christian era, in which sky-ground dualisms form a predominant theme, liked in numerous convoluted ways to the issue of resurrection and immortality of the soul. These writings, the "Hermetic texts", were believed to have been the work of the ancient Egyptian wisdom god Thoth (known to the Greeks as Hermes), who in one representative passage makes the following remarks to his disciple Asclepius: "Do you not know, Asclepius, that Egypt is an image of heaven? Or to speak more exactly, in Egypt *all the operations of the powers which rule and work in heaven have been transferred down to earth below?*"[40] The purpose to which these powers were harnessed, in the Hermetic view, was to facilitate the initiate's quest for immortality.
>
> Curiously, precisely such a quest for precisely such a goal – "a life of millions of years" – is spelled out in ancient Egyptian funery texts which supposedly pre-date the Hermetic writings by thousands of years. In one of these texts, *Shat Ent Am Duat* – the *Book of What is in the Duat* – we find what appears to be an explicit instruction to the initiate to build a replica on the ground of a special area of the sky known as the "hidden circle of the Duat": "Whosoever shall make an exact copy of these forms... and shall know it, shall be a spirit and well equipped both in heaven and earth, unfailingly, and regularly and eternally."[41]
>
> Elsewhere in the same text we hear again of "the hidden circle in the Duat...in the body of the Nut(the sky)": "Whosoever shall make *a*

---

[39] Robert Bauvall and Graham Hancock, *The Message of the Sphinx* (New York: Three Rivers Press, 1996), p. 78.

[40] Ibid., p. 78, citing *Hermetica*, transl. Sir Walter Scott (Boston: Shambala, 1993), *Asclepius III: 24b*, p. 341, emphasis added. Scott notes the Latin *translatio*, which he has translated as "transferred", probably translates the Greek word μεταθεσις.

[41] Ibid., citing the eleventh division of the *Duat*, in the "Book of What is in the Duat", transl. Sir E.A. Wallis Budge, *The Egyptian Heaven and Hell* (London: Martin Hopkinson and Co., 1925), p. 240.

copy thereof... it shall act as a magical protector for him both in heaven and upon earth."[42],[43]

Several points must be noted both about the ancient texts these authors have cited as well as about the conclusions they have drawn from them:

(1) the original Latin[44] reads *"An ignorans, O Asclepi, quod Aegyptus imago sit caeli, aut, quod verius, {...} translatio aut decensio omnium quae gubernatur atque exercentur in caelo?"*[45] The English phrase "all the operations of the powers" has therefore been supplied by Scott in his translation to compensate for a lacuna due to the deteriorated condition of the manuscript.

(2) Moreover, as Hancock and Bauvall observe, the purpose of this was "to facilitate the initiate's quest for immortality."[46] They do not consider the hypothesis, suggested by this and the other quotations they cite in their work, that these texts preserve the basic principles, in a very detailed manner, of a very advanced paleophysics – even by contemporary standards – and a resulting sophistication of technological application deriving from it, in the only language a less sophisticated culture could communicate them in: the language of religion and metaphysics.

(3) The combined implication of the three texts cited by Bauvall and Hancock is of some sort of military application; the third text cited speaks particularly of a

---

[42] Ibid., the Twelfth division of the *Duat*, p. 258.

[43] Bauvall and Hancock, op. cit., pp. 78-79.

[44] The *Asclepius* is preserved only in a Latin translation of a Greek original no longer extant, which in turn purports to be the translation from ancient Egyptian texts. By "original" here I denote only the language from which the English of Hancock and Bauvall's quotation has been translated.

[45] *Asclepius 24b, Hermetica,* Sir Walter Scott, op. cit., p. 340.

[46] Bauvall and Hancock, op. cit., p. 78.

"magical" protection both in heaven and on earth, suggesting at least in part that the vast engineering of the Giza complex was at the minimum *defensive* in nature. It should also be noted that the physics involved is *on a celestial and planetary scale.*

(4) Viewing these authors' cited texts from the standpoint of the physics of harmonic entanglement of systems outlined in the next chapter leads to different conclusions:

(a) the purpose of the "as above, so below" engineering is to couple the energy of the motions of the Milky Way galactic system to the earth, in short, the Giza complex is an *open* system, a "coupled oscillator" (and as will be seen, interferometer) to the motions of the galaxy. In short, Giza embodies the physics of open harmonically entangled systems.

(b) The resulting entanglement of open systems engineering represents a reliance upon a very different conception of energy within that paleophysics, wherein energy is the *result* of systems that are entangled, and *where the aether itself is not, as for classical physics, a passive inert "medium", but an active transmutative medium of entanglement.* That is, energy is the result of the "information in the field" of space[47] itself, and therefore of the space engineered at Giza. This energy is as "unfailing, regular, and eternal," as the motion of the celestial bodies it harmonically duplicates.

Hancock and Bauvall go on to cite yet another ancient text that indicates a military purpose behind the architecture and engineering at Giza (and again, they ignore this very clear indication):

---

[47] "Space" here may also be called the "Zero Point" or "vacuum", provided the meaning of "voidness" is not taken to be present with the term.

> The Shabaka texts tell us how the god was taken and buried "in the land of Sokar":

> "This is the land... the burial [place] of Osiris in the House of Sokar... Horus speaks to Isis and [her sister] Nephthys: "Hurry, grasp him..." Isis and Nephthys speak to Osiris: "We come, we take you..." They heeded in time and brought him to Land. He entered into the hidden portals... of the Lords of Eternity. *Thus Osiris came into the earth, at the Royal fortress, to the north of the land to which he had come.*"[48]

This important clue is ignored by the authors, though it corresponds not only to the implications of texts that they have cited previously, as well as to the wider body of paleographic evidence suggesting the existence of a sophisticated physics and its weaponization at Giza.

The *Asclepius* of the *Hermetica*, with which Baulvall and Hancock began their considerations in the quotation cited at the beginning of this section, contains a wealth of clues regarding the principles of ancient paleophysics. Certain features of this ancient paleophysics are propositions of a purely philosophical and metaphysical nature, i.e., concern the basic theoretical assumptions of that paleophysics. Other clues are purely "scientific" and do not concern themselves with outlining the paradigms of the theoretical model of that paleophysics. However, the problem is that both types of statements are couched in the metaphysical and philosophical language of a legacy civilization, and thus the two types of statements are difficult to distinguish. There is, moreover, a third class of statement, couched in the typical pantheistic or pan*en*theistic language of such texts, that is usually misunderstood as belonging to the first class of statement when in fact it belongs to the second class, for as has been shown, crucial to the model of this ancient paleophysics is the view of the universe as a "living" thing, i.e., as a supersystem of entangled subsystems, *much like an organism.*

---

[48] Miriam Lichtheim, *Ancient Egyptian Literature* (Berkeley: University of California Press, 1975), Vol I, p. 53, cited in Bauvall and Hancock, op. cit., p. 144, emphasis added.

The *Asclepius* and the *Libellus* of the *Hermetica* have been selected for this examination, not because they are the only such texts that embody the principles of this paleophysics, but rather because they are relatively representative of such texts and more or less readily available. In order to facilitate an accurate perception of this paleophysics and of the interpretive principles involved in any such examination of ancient texts, initially each text examined will be given its own section heading, which heading states the principle involved in its original metaphysical language, followed by a restatement of the same principle in terms of the language of physics, cosmology, and systems theory.[49]

<u>*(1)*</u>
Soul as the All-Pervasive Substance of the Cosmos:
The Continuum, Vacuum, or Void Contains
Information in the Field, By Means of Which All Systems are, or
may be, Entangled:
Asclepius I:2b; 3c:

> This whole, then, which is made up of all things, or is all things, consists, as you have heard me say before, of *soul and corporeal substance.* Soul and corporeal substance together are embraced by nature, and are by nature's working kept in *movement; and by this movement, the manifold qualities of all things that take shape are made to differ among themselves, in such sort that there come into existence individual things of infinitely numerous forms,* by reasons of the differences of their qualities, *and yet all individuals are united to the whole;* so that we see that.... *Matter is one, soul is one, and God is one.*[50]

> For by all the heavenly bodies... there is poured into all matter an uninterrupted stream of soul.[51]

---

[49] Once again, the basic historiographical presupposition underlying this method is that there was an ancient Very High Civilization of such technical and scientific sophistication that it surpassed even contemporary civilization in those respects, and that the ancient civilizations of the Indus Valley, Mesopotamia, Egypt and so on are its considerably *declined* legacies.

[50] *Asclepius I:2b,* Scott, op. cit., pp. 289, 291, emphasis added.

[51] Ibid., I:3c, p. 291.

(1) This passage closely parallel some views within modern physical cosmology that regard the universe as a network of interlocked complex systems. In such a view, a local disturbance resonates throughout the whole system, in much the same way that a local disturbance in the body of a living organism affects the whole organism. Hence this view, as expressed by the "legacy civilizations" has been profoundly misinterpreted: such civilizations understood the metaphor *literally*: the universe *becomes* a living entity, suffused with "soul" to *account for* the phenomenon of systems entanglement.

(2) Thus, the following types of phrases become the way for legacy civlizations to speak easily about a physics they no longer are capable of understanding:

   (a) "Soul and matter are one" refer to non-local, entangled systems, "soul" being the "information in the field" or the condition of the entanglement of such non-local systems. One may think of it simply as the spatial arrangement of the components of the system, as its *geometry*.

   (b) "Motion" thus gives rise to differentiated systems;[52] "soul" as "the information in the field" can thus also be viewed as the underlying substrate or sub-quantum fluctuations in the aether, the stressing of which produces non-local entangled systems. They are entangled because they arise by similar processes from the same *Urstoff,* or "early stuff", the "stuff" that existed *before* matter

---

[52] In one respect, at least, even the legacy civilizations were considerably more sophisticated than modern theoretical physics in their insistence that *everything* is in motion. Much of modern pre-relativistic physics assumed that local spacetime is flat and not curved, which to the ancients, particularly when stellar or celestial motions were in view, would have appeared to be so much nonsense since the theoretical model was in flat contradiction – to coin a pun – to observation.

gathered into atoms, stars, solar systems, galaxies, and galactic clusters and so on.

(c) This soul-matter aether or *Urstoff* is undifferentiated space-time and thus represents a vast potential of information in the field. It is, in short, a space-time *devoid* of the geometric configuration brought to bear on it by various physical systems.

(d) Since motion is the primary means by which the "variety of forms" i.e., non-local entangled systems, come into existence, their various motions constitute the *basis* of their entangling; that is, *since the motion of various systems is in view, then time, and not forces, mass, or any other such entity, is the primary differential – or "thing in view" - of this ancient paleophysics.*

(e) The entangling of non-local systems is accomplished by some celestial component ("..for by all the heavenly bodies..."). This is closely parallel to the view of plasma cosmology for even the type of motion is suggested by it: the rotational vorticular motion of electromagnetic vortexes, found from the largest galaxies down to the smallest plasma phenomena, and thus preserving a kind of symmetry across degrees or gauges of scale and size. A physicist will perhaps appreciate the significance of this more readily than a layman. A modern physicist tends to think of reality in terms of a series of physical laws that only apply to items of a certain size. For example, the laws of quantum mechanics work very well with atomic and sub-atomic particles, but at the scale of planets and stars and galaxies, not so well. For items of that size, a different set of laws appears to hold true. The quest for a unified series of physical laws is one of the holy Grails of physics. But this would appear to be precisely what may have once existed, for the texts indicate that the paleoancients apparently viewed *all* physical objects in more or less the same way.

(2)
The Mind-Aether Connection:
An Anthropic Cosmological Principle:

The first class of purely metaphysical statement of a cardinal theoretical assumption of this ancient paleophysics is represented by a short statement in the *Asclepius I: 6b:*

> Mind, a fifth component part, which comes from the aether, has been bestowed on man alone; and of all beings that have soul, man is the only one whose faculty of cognition is, by this gift of mind, so strengthened, elevated, and exalted, that he can attain to knowledge of the truth concerning God.[53]

The presuppositions contained in this passage provide a basis for understanding some of the theoretical assumptions of this paleophysics:

(1) God is intelligence and can "engineer" the vacuum aether to bring forth the variety of the cosmos;

(2) The universe is thus the product of intelligence and contains design, i.e., evidence of that intelligence;

(3) The link between that aether and mind itself is immediate, and only man has a rational faculty to perceive the engineering of the vacuum implied by that immediacy;

(4) The universe is thus intelligible to man, which is the ancient paleophysical embodiment of the modern "anthropic principle" in physics;[54]

---

[53] *Asclepius I: 6b,* Scott, op. cit., p. 297.
[54] Cf. John D. Barrow and Frank J. Tipler, *The Anthropic Cosmological Principle* (Oxford, 1988). For the contrary view that metaphysical or religious principles are an impediment to cosmology, and the relationship of religion to

(5) "Mind" thus connotes the "information in the field" or the configuration of various interlocked systems; and finally,

(6) If God can engineer the vacuum in such a way as to give rise to entangled non-local systems, then man, who possesses an analogous rational faculty, can likewise engineer the vacuum, and on a similarly grand scale:

> *God, the Master of eternity, is first; the Kosmos is second; Man is third. God, the Maker of the Kosmos and of all things that are therein, governs all things, but has made Man as a composite being to govern in conjunction with him. And if man takes upon him in all its fullness the function assigned to him... he becomes the means of right order to the Kosmos, and the Kosmos to him; so that it seems the Kosmos (that is, the ordered universe) has been rightly so named, because man's composite nature has been ordered by God.*[55]

Physicists will recognize this immediately as being analogous to the self-selection principle operative in the Anthropic Cosmological Principle:

> This approach to evaluating unusual features of our Universe first re-emerges in modern times in a paper of Whitrow who, in 1955, sought an answer to the question *'why does space have three dimensions?'*. Although unable to explain why space actually has, (or perhaps even why it must have), three dimensions, Whitrow argued that this feature of the World is not unrelated to our own existence as observers of it. When formulated in three dimensions, mathematical physics possesses many unique properties that are necessary prerequisites for the existence of rational information-processing and 'observers' similar to ourselves.
>
> ....
>
> Our definition of the (Weak Anthropic Principle) is motivated in part by these insights together with later, rather similar dieas of Dicke who, in 1957, pointed out that the nmber of particles in the observable extent of the Universe, and the existence of Dirac's famous Large Number

---

physical cosmology, cf. Eric J. Lerner, *The Big Band Never Happened* (Vintage), 1992. Suffice it to say that the ancient paleophysics' cosmology more closely resembles that of the plasma cosmology of Hannes Alfven that Lerner outlines, but nonetheless incorporated into a metaphysical structure.

[55] *Asclepius I: 10*, Scott, op. cit., p. 305.

Coincidences '*were not random but conditioned by biological factors*'. This motivates the following definition:

Weeak Anthropic Principle (WAP): The observed valus of all physical and cosmological quantities are not euqally probably but they take on values restricted by the requirement that there exist sites where carbon-based life can evolve and by the requirement that the universe be old enough fot it to have already done so.

Again we should stress that this statement.... Expresses only the fact that those properties of the Universe we are able to discern are self-selected by the fact that they must be consistent with out own evolution and present existence.[56]

This Anthropic cosmology must be further modified to account for the results of Quantum mechanics. Barrow and Tipler point out that the physicist Wheeler was thus led to formulate a version of the Anthropic principle called the "Participatory Anthropic Principle" or PAP which may be succinctly stated as follows: "*Observers are necessary to bring the Universe into being.*"[57] Inclusion of the theoretical models of systems theory and information theory result in a Final Anthropic principle, which it will be observed, comes the closest to the principle as embodied in the *Hermetica*: "*Intelligent information-processing must come into existence in the Universe, and, once it comes into existence, it will never die out.*"[58] Moreover, the passage of the *Hermetica* thus fulfills one of LaViolette's conditions for the examination of such texts for possible paleophysical content, namely, that such texts preserve the same *sequential* ordering of events: first the universe (Kosmos), then the observer (man). One also discovers in a latter passage (cf. section (3a) below) another profound allusion to the modern Anthropic Cosmological Principle in that space is a necessary condition of existence.

---

[56] Barrow and Tipler, op. cit., pp. 15-16, emphasis in the original.

[57] Ibid., p. 22, emphasis in the original.

[58] Ibid., p. 23, emphasis in the original.

(3)
The Science of Music and "How to Do It":
The Harmonic Specification and Entanglement of Diverse systems:
Asclepius I: 13-14a:

Belonging definitely to the third class of statements is the *Asclepius I: 13-14a* which contains a statement of the ancient doctrine of the cosmic symphony or "harmony of the spheres" (κοσμικη συμφωνια):

> *And to know the science of music is nothing else than this – to know how all things are ordered, and how God's design has assigned each its place; for the ordered skill in which each and all... are wrought together into a single whole yields a divinely musical harmony...*[59]

That is to say, the method whereby discrete non-local systems arise, and by which they may therefore be entangled, is via their harmonic relationships. And once we say "harmonics", we are over course back to the primary differential, or "thing in view", time.

This point cannot possibly be lingered over too long, for it suggests a method of verification of the physics being elucidated, and it is worth repeating this verificational possibility: *The vacuum or ZPE potential of any given point or system must be so engineered to be in consonance with the base planetary system and any entangled (solar and galactic) systems in which that planetary base resides.*[60] Additionally, this principle serves to explain why attempts to verify various zero point energy experiments often fail, for the structured potential of the vacuum has *not* been harmonically "tuned" or geometrically duplicated.[61] Further

---

[59] *Asclepius I: 13-14a*, Scott, op. cit., p. 311, emphasis added.

[60] "Base planetary, solar, and galactic" systems are defined in chapter six. ZPE is simply the physicist's abbreviation for "Zero Point Energy".

[61] Tesla often remarked on this phenomenon. Cf. Gerry Vassilatos, *Secrets of Cold War Technology: Project HAARP and Beyond* (Bayside, California: Borderland Sciences, 19), pp. 38, 42, 45 66. Lt. Col. Thomas E. Bearden has observed a similar phenomenon, and this may in part also explain the mixed

methods of verification will be addressed subsequently. One may summarize this principle by saying that matter exists in a "primordial" or ground state of pure potential information in the vacuum field, since the harmonic stressing of that field accounts for the variety of forms.[62]

(a)
"Primordial Matter":
Plasma Cosmology, Open Systems, and Space as an
Antecedent Condition of Being:
Asclepius II: 14-15:

Two conceptions fundamental to the ancient paleophysics – matter and space – now merge to form what at first glance appears to be a classical definition of the *aether lumeniferous* of nineteenth century physics: a formless (i.e., undifferentiated) dimensionless (or in some versions, infinitely extended) substance that pervaded all things. This resemblance is only superficial, however, for the aether of nineteenth century physics was a *passive* medium, whereas the aether of the ancient paleophysics was often described as "fertile" or "fecund", that is, was an *active, transmutative, and creative medium.*

> Matter, though it is manifestly ungenerated, yet has in itself from the first the power of generating; for an original fecundity is inherent in the properties of matter.... Matter then is generative by itself, without the help of anything else. It undoubtedly contains in itself the power of generating all things.
> .... Thus the space in which is contained the universe with all things that are therein is manifestly ungenerated.... For the existence of all things that are would have been impossible, if space had not existed as an antecedent condition of their being.[63]

---

results in verifying endothermic fusion claims. Cf. Thomas E. Bearden, *Gravitobiology* (Tesla Books, 1991), pp. 91 – 92, n. 73.

[62] Cf. *Asclepius II: 14b*, Scott, op. cit., p. 311.

[63] *Asclepius II: 14-15*, Ibid., p. 313.

Another profound indicator of an ancient paleophysics in this passage is that space and matter are both said to be "ingenerate", though conceptually for the theoretical model itself, they are not quite the same thing. Matter, while ingenerate (i.e., in itself formless and undifferentiated), can generate discrete forms or systems via harmonic stressing, space on the other hand, while ingenerate, does *not* generate but is that *in* which matter exists and generates. *Viewed together, matter and space comprise the ancient conception of an active transmutative aether.*

(b)
"Primordial Matter":
Transmutative Aether as the Potential of All Forms or Systems:
Asclepius III:17a:

As a consequence of these considerations, matter may be viewed from a slightly different perspective, as the harmonic or "overtone series" of all potential forms or systems: "Matter is the recipient of all forms; and the changes and unbroken successions of the forms (are wrought by means of Spirit)."[64] The reference to the "changes of forms" being "wrought by "Spirit" may thus be interpreted as implying that primordial matter may be viewed as containing "all potential information in the field." The theoretical basis for the entangling non-local systems via harmonic resonances existing between them has now been laid in the principles outlined above in (3), (3a), and (3b).

As a further consequence of these postulates, it follows that no such thing as an "absolute vacuum" or "void" exists, because such a notion would not allow for the entanglement of *any* non-local systems.[65] "I hold that no such thing as void exists, or can have existed in the past, or ever will exist."[66] This too is an unusual

---

[64] *Asclepius III: 17a,* Ibid., p. 317.

[65] In modern physics terms, a solution for the quantum wave equations that would yield the Einstein-Podolsky-Rosen effect could not exist if a "void-vacuum" existed, since such absolute space would have no geometric characteristics.

[66] *Asclepius III: 33b,* Scott, op. cit., p. 319.

twist on the Anthropic Principle of contemporary physics, for it is being made in conjunction, not with Big Bang cosmologies which *do* presuppose such a "void-vacuum" (for as will be seen subsequently, Michelson-Morley "conclusively" dispatched of the aether wind!), but in conjunction with a model more akin to the plasma cosmology or string theory and M-Theory of modern theoretical physics. Hermes maintains that this is so because

> The thing that seems void, *however small it may be*, cannot possibly be void of spirit and of air.
> *And the like must be said of space.* The word 'space' is unmeaning when it stands alone; for it is only by regarding[67] something which is in space, that we come to see what space is.[68]

This may be one of the most significant paleophysical principles in evidence in all the ancient literature, and its bold contemporary theoretical significance must be stated plainly: *The very act of observing space means that it is not the inert passive aether lumeniferous of nineteenth century physics or void of contemporary physics, but means that it must* **contain**, *as a bare minimum, "information in the field" or material of "sub-quantum" size.* Again, the word "spirit" here is taken to indicate this "information potential of the field."

<u>*(4)*</u>
"Destiny" or "Fate":
The Notion of "Time Locks" or the "Base Time" of the Base
Planetary System:
Asclepius III: 27b; 35:

With these conceptions in mind, the most difficult and egregiously problematic of the doctrines of the ancient paleophysics may now be approached, and hopefully, appreciated for what it really is: a profound insight into the nature of the relationship of time to the harmonics of entangled non-local

---

[67] Observing!

[68] *Asclepius III:34a*, Scott, op. cit., p. 321.

systems. Briefly, this insight is mediated, and profoundly misunderstood, by the legacy civilizations' astrological doctrine that an individual's "destiny" is "written in the stars", i.e., that it is to some extent influenced by celestial mechanics, or the "star sign" of the zodiacal house an individual is born under:

> The Ruler of the Decani – that is, the thiry-six fixed stars which are called Horoscopi – is the god named Pantomorphos;[69] he it is that gives to the individuals of each kind their diverse forms.
>
> The seven spheres, as they are called, have as their Ruler the deity called Fortune or Destiny, who changes all things according to the law of natural growth, working with a fixity which is immutable, and which is yet varied by everlasting movement.[70]

If one now suspends focus on the "astrological" component and reads the quotation merely from the viewpoint of systems harmonic entanglement, a rather astonishing set of postulates of the ancient paleophysics emerges:

(1) That at the point that undifferentiated primordial matter generates any distinct form, motion occurs, and therefore the motion of that system is peculiar to it, i.e., every physical object without exception resonates to some frequency, and has its own series of harmonics;

(2) Each system therefore bears is own motion;

(3) Each system is therefore related to every other such system by dint of the harmonics of such motions; and therefore,

(4) Each system bears its own "time reference" or "base time" to other such systems, expressed as a relations of time when a system emerges to systems already in existence.

In short, time is the fundamental organizing component of systems and *the key determinant* both in their potential harmonic

---

[69] The Greek name means "All Forms".

[70] *Asclepius III: 27b;* Scott, op. cit., p. 325.

entanglement and therefore in any attempt to entangle non-local systems artificially.[71]

> For it is impossible that any single form should come into being which is exactly like a second, if they originate at different points of time, and at places differently situated; but the forms change at every moment in each hour of the revolution of that celestial circle in which resides that god whom we have named Pantomorphos. Thus the type persists unchanged, but generates at successive instants copies of itself as numerous and different as are the moments in the revolution of the sphere of heaven....[72]

That is to say, given two systems that in all other respects are identical, and that arise at different "base times" (and therefore places), the "time differential" between them suffices to distinguish them. One may even speculate what this may mean. If one envisions two identical life forms existing in widely separated planetary systems, it is probable that the average life spans of these forms will vary as a function of the various motions of those systems relative to the larger galactic system in which the planetary systems exist, even if they share a common galactic system. Something like this, of course, is predicted by General Relativity for systems approaching the velocity of light. But what is being specifically maintained by the ancient paleophysics – albeit in astrological form handed down by the legacy civilizations – is that these effects occur in and are significant for non-relativistic velocities as a consequence not of the motion of such individual systems as such, *but as a consequence of motions of systems in certain (harmonic) relationships to other systems.*

A further consideration emerges from these speculations. There is a minor "discussion" of sorts occurring in current cosmology. This discussion seeks to answer the question: "What type of civilization would be capable of genuine interplanetary, interstellar, or 'time' travel?" An elaborate hypothesis has been constructed to answer this question. This hypothesis posits

---

[71] This point will be elaborated in chapter six.

[72] *Asclepius III:35,* Scott, op. cit., p. 329.

different "civilization types" on the basis of their knowledge of certain physics principles and their ability to manipulate or engineer certain kinds of energy.

> Astronomer Nikolai Kardashev of the former Soviet Union once categorized future civilizations in the following way.
>
> A Type I civilization is one that controls the energy resources of an entire planet. This civilization can control the weather, prevent earthquakes, mine deep in the earth's crust, and harvest the oceans. This civilization has already completed the exploration of its solar system.
>
> A Type II civilization is one that controls the power of the sun itself. This does not mean passively harnessing solar energy; this civilization mines the sun. The energy needs of this civilization are so large that it directly consumes the power of the sun to drive its machines. This civilization will begin the colonization of local star systems.
>
> A Type III civilization is one that controls the power of an entire galaxy. For power sources, it harnesses the power of billions of star systems. It has probably mastered Einstein's equations and can manipulate space-time as well.
>
> The basis of this classification is rather simple: Each level is categorized on the basis of the power source that energizes the civilization.[73]

It will be noted that this model of civilization types is dependent upon several implicit assumptions necessary to the cosmology and physics assumed in the scheme. As these assumptions will be critiqued in the next chapter, they will not be mentioned here, save that a conclusion inevitably follows from this scheme of civilization types. Such civilizations would be incomparably older than our own to allow for the longer time such technologies would require for development.

However, there are a number of flaws to this view (and to the cosmology and physics implicit within it):

    (1) Such extrapolations are based largely on a "linear" notion of scientific progress and the evolution of life, a

---

[73] Michio Kaku, *Hyperspace: A Scientific Odessy through Parallel Universes, Time Warps, and the 10th Dimension.* (Oxford: Oxford University Press, 1994), pp. 277-278.

view which is problematical, since there is no *a priori* reason why a "scientific revolution" could not occur that would propel such a civilization to accomplish these achievements in a time frame much shorter by several orders of magnitude;

(2) Such extrapolations are based upon a fundamentally *human* basis of experience of the accumulation of knowledge over time from one generation to the next. But if one can imagine an intelligent species of life with an average life span of one thousand of our years, then the amount of knowledge will accumulate and transmit from generation to generation at a much more rapid rate.

Indeed, something like the second proposition is implied by most paleographic evidence regarding the ancient Very High Civilization. Most ancient legacy civilizations preserve some form of tradition that indicates that the average life span of the members of that civilization *was much longer*, and this would permit the rapid accumulation of knowledge.

Finally, it is to be noted that the *physics* of these civilization types is *not* based upon the harmonic systems entanglement being postulated here. However, it is curious that the three types of civilizations modeled by Kardashev are precisely the three base systems whose energy is explicitly being tapped at Giza, as indicated by the paleographic evidence. *In short, in Kardashev's classification scheme, the civilization that built the Great Pyramid was a Type III civilization!*

<u>(5)</u>
"As Above, So Below:"
<u>Galactic and Terrestrial Systems Entanglement:</u>
<u>Asclepius III:24b:</u>

The quotation cited by authors Bauvall and Hancock that began this section contains the first detailed clue to the application of the physics of harmonic systems entanglement at the Giza compound:

> Do you not know, Asclepius, *that in Egypt is an image of heaven, or to speak more exactly, in Egypt all the operations and powers which rule and work in Heaven have been transferred to earth below?* Nay, it should rather be said that the whole Kosmos dwells in this our land as its sanctuary.[74]

It is this text which has been the basis of so much fruitful research into the function of the Great Pyramid and the Giza complex. Unfortunately, its main message, however, is lost in the static of the Observatory and "extra-terrestrial" hypothesis: The Great Pyramid was a machine of a "Type III civilization" that drew upon the power of the entire galaxy.[75]

(a)
Kosmos as Contained in Time:
Time is The Primary Differential of the Ancient Paleophysics:
Asclepius III:30

The harmonic entanglement of non-local systems implies a primary focus on the movements of systems and the relationship of these movements to each other (harmonics):

> Kosmos is that in which time is contained; and it is by the progress and movement of time that life is maintained in the Kosmos. The progress of time is regulated by a fixed order.... All things being subject to this process, there is nothing that stands fast, nothing fixed, nothing free from change, among the things which come into being, neither among those in heaven nor among those on earth.[76]

Several important elements of the paleophysics must be noted:

---

[74] *Asclepius III: 24b,* Scott, op. cit., p. 341.

[75] Actually, the civilization Types are a misnomer in this physics model, since the essence of this type of systems physics is the drawing of power from all three sources – planetary, solar, and galactic – simultaneously by coupling them together. Perhaps one should speak of it as a Type *IV* civilization.

[76] *Asclepius III: 30,* Scott, op. cit., p. 351.

(1) since Kosmos itself connotes the harmonic order existing between various systems, then motion, change, and hence *time* is the primary component of that harmony;

(2) Time is an ordered process, and therefore,

(3) Time is the primary differential of that ancient paleophysics.

The last point may not fully be appreciated unless one take it to its full extent: imagine recasting all the equations of physics – motion, force, pressure, mass, and so on – with time being the primary differential or "concept in view", and not such concepts as gravitational constant, magnetic moment, and so on.

(b)
Destiny and Necessity:
Time as a Force and the Reinterpretation of "Force":
Asclepius III: 39:

If time is the primary differential of this ancient paleophysics, then:

> That which we call Destiny, Asclepius, is the force by which all events are brought to pass; for all events are bound together in a never broken chain by the bonds of necessity. Destiny then is either God himself, or else it is the force which is next after God...[77]

If time is the primary differential, the distinctions of other physical objects forces – gravity, electromagnetic, strong and weak nuclear forces - would expressible as "time differentials", since such forces are the result of the geometry of the systems themselves.[78]

---

[77] *Asclepius III:39*, scott, op. cit., p. 363.

[78] This rather obscure point will (hopefully) be made more clear in the next chapter.

(c)
Downward and Upward motion:
Entropic and Non-Entropic Motion:
Libellus I:4:

> And in a little while, there had come to be in one part a downward tending darkness, terrible and grim.... And thereafter I saw the darkness changing into a watery substance..."[79]

More will be said in due course on the twin concepts within ancient dualisms of downward-upward motions and their relationship to darkness and light. In terms of their possible symbolism of the ancient paleophysics, however, these dialectical polarities denote the tendencies of the laws of entropy and of anti-entropic, self-organizing systems on the other, exactly the type of modeling one would expect with the non-equilibrium thermodynamics of reaction-diffusion systems to be encountered later.

(6)
Life as the Union of Word and Mind:
Potential and Realized Information in the Field:
Libellus I:6:

The advanced paleophysics of necessity not only viewed the universe as an open complex system comprised of several complex sub-systems -- hence giving rise to the metaphor of the universe as an organism in the legacy civilizations – but it also inevitably had implications for the view of life itself as intelligently constructed: "They are not separate one from the other: for life is the union of Word and Mind."[80] That is, if "mind" is the "potential information in the field", "Word" is a particular manifestation of it. More importantly, there is an implicit Anthropic Cosmology here, for if life is intelligently constructed, it is intelligible, and therefore requires an intelligent observer to observe it.

---

[79] *Libellus I:4*, Ibid., p. 115.
[80] *Libellus I:6*, Ibid., p. 14.

"Light" as "Innumerable Powers" and a "World without
Bounds":
The Electromagnetic Spectrum and an Infinite Unbounded
Universe:
Libellus I:7-8a:

A crucial component of the ancient paleophysics, and one
which is again being advocated with increasing scientific rigor
within modern physical cosmology, is the primacy of
electromagnetic phenomena:[81] "I saw in my mind that the light
consisted of innumerable powers, and had come to be an ordered
world, but a world without bounds."[82] Simply put, one has here
the essence of the plasma cosmology of Swedish physicist Hannes
Alfvén: the world was formed out of electromagnetic fluctuations
that can occur even in "vacuum" space.

Not only do the multiplicity of forms in the universe result
from variable structuring of the "vacuum" and from systems
entanglement, but this passage also provides a clue into how the
ancient paleophysics interpreted cosmological origins and affords
an important glimpse into its technology:

(1) In contrast to the post-relativistic "Big Bang"
cosmology with its finite and unbounded universe, or
its more recent "many universes" quantum mechanical
models, the ancient cosmology viewed the universe as
infinite and unbounded both spatially *and temporally;*

(2) Moreover, the universe was not, on the "paleoancient" view, *isotropic,* i.e., its model did *not* assume,
even for theoretical simplicity or ease of mathematical
modeling, that it was initially of uniform consistent

---

[81] Cf. Hannes Alfvén, "Cosmology in the Plasma Universe," *Laser and
Particle Beams,* vol. 16 (Aug-1988), pp. 389-398; *Cosmic Plasma* (Holland: D.
Reidel, 1981); Eric J. Lerner, *The Big Bang Never Happened* (Vintage, 1991).

[82] *Libellus I: 7-8a,* Scott, op. cit., p. 117.

composition at all times and places; it was, rather, *anisotropic,* matter was distributed unequally throughout it;

(3) The term "light" and the rest of the quotation suggest a sophisticated knowledge of electro-magnetic phenomena, consisting of various "powers" or properties of a certain sort inhering in certain frequencies and modulations;

(4) There is some sort of connection between points (1) and (2) on the one hand, and point (3) on the other, since they are contextually linked. One may speculate that the original reason for this linkage was lost in the text due to the legacy civilizations not possessing a religious metaphor capable of conveying the technicalities of the linkage.[83] Equally possible, of course, is that the original paleoancient Very High Civilization did not transmit that knowledge, for whatever reason, or that it was deliberately encoded by it, or lost through normal processes of textual corruption.

(8)
"Orbits of Administrators":
Base Solar System and the Cosmology of No Temporal
Beginnings:
Libellus I:11a:

The then seven known planets, or "administrators", were then "worked" by Mind and Word, the conditions that constitute life:

And Mind the Maker worked together with the Word, and encompassing the orbits of the Administrators, and whirling them round with a rushing movement, set circling the bodies he had made,

---

[83] This point will emerge with even more force in the subsequent examination of ancient Sanskrit texts.

and let them revolve, travelling from no fixed starting point to no determined goal....[84]

There are several noteworthy features of the ancient paleophysics' cosmology contained here, features which bear a striking resemblance, once again, to Alfvén's plasma cosmology:

(1) the universe is unbounded spatially and temporally; there is no initiatory "Big Bang"; rather, electromagnetic vorticular rotation is a basic cosmological principle;

(2) the arrangement of orbits of the planets in the solar system are exact enough to support life; since life is the union of Mind and Word; where such conditions are present, life is likely;

(3) The text indicates that the planets not only revolve in the orbits, but upon their axes;

(4) Finally, and most importantly, the text indicates the coupling of the earth (base planetary system) to the planets and sun (base solar system).

(9)
"Man's Station in the Maker's Sphere":
The Anthropic Cosmological Principle of Paleophysics and The
Anentropic Principle:
Libellus I: 13a-14:

The idea that life itself is "the union of Mind and Word" implies moreover that the physical structure of the cosmos is itself so entangled that man himself cannot only discover the principles of its architecture, and engineer them himself, but also that he plays a crucial role in the anti-entropic tendencies of the universe:

> And man took station in the Maker's sphere, and observed the things made by (Mind the Maker), who was set over the region of fire; and having observed the Maker's creation in the region of fire, he willed to

---

[84] *Libellus I: 11a*, Scott, op. cit., p. 119.

make things for his own part also; and his father gave permission....
Having in himself all the working of the administrators; and the
administrators took delight in him, and each of them gave him a share
of his own nature.

And having learnt to know the being of the administrators, and
received a share of their nature, he willed to break through the
bounding circle of their orbits; and he looked down through the
structure of the heavens, having broken through the sphere, and showed
to downward-tending nature the beautiful form of God
(και εδειξε τη κατωφερει φυσει την καλην του Θεου μορφην).[85]

A clearer statement of the Participatory Anthropic Cosmological
Principle there could not be.

(10)
"Extension" and the Distinction Between God and Space:
Kosmos as a Unified, Complex System of entangled Sub-Systems:
Space, Object, System, Motion, and Opposition:
Libellus II:1-6b:

A very rich passage for paleophysical concepts may be found
in the *Libellus II: 1-6b:*

> Of what magnitude then must be the Space in which the Kosmos is
> moved? And of what nature? Must not that Space be far greater, that it
> may be able to contain the continuous motion of the Kosmos, and that
> the thing moved may not be cramped through want of room, and cease
> to move? --*Ascl.* Great indeed must be that Space, Trismegistus. –
> *Herm.* And of what nature must it be, Asclepius? Must it not be of
> opposite nature to the Kosmos? And of opposite nature to body is the
> incorporeal. –*Ascl.* Agreed. –*Herm.* That Space then is incorporeal....
> Space is an object of thought, but not in the same sense that God is, for
> God is an object of though primarily to himself, but Space is an object
> of thought to us, not to itself.[86]

---

[85] *Libellus I: 13a-14,* Scott, op. cit., p. 121.
[86] *Libellus II:1-6b,* Ibid., pp. 135, 137. For the connection between the
notions of God and space via the concept of extension, cf. the debates between
Descartes and Newton.

Contrary to the view of Newtonian mechanics, for the ancient paleophysics there is a distinction between God and Space, a distinction made on the basis of the following dialectic of oppositions:

God	*is not*	Space
Is known		Is not known
To Himself		to itself.

That is, the principle distinction between the two is that Space requires *an observer*.[87]

The third entity, Kosmos, which is to be understood as a "material, corporeal, complex system" is further distinguished from God and space by the same process of dialectical opposition:

God	*is not*	Space
(a) known to		(a) not known
to himself		to itself
(b) unmoved		(b) unmoved
(c) incorporeal		(c) incorporeal

Kosmos
(a) not known to itself
(b) moved
(c) corporeal

Space, then is the universal prerequisite condition for motion, and while - pace Einstein - it does not move, it may be "bent" or

---

[87] Hence the flaw in Newtonian mechanics, for once the observer was left out of the mathematical model, God and Space appeared to be the same entity based upon common properties involving extension.

warped. Conversely, the material universe's one invariable condition is its change, variance, or motion.

(a)
Every Living Thing is Moved from Within:
Motion is the result of the Potential of the information within the Field:
Libellus: II:8b, 10:

> The movement of the Kosmos, then, and of every living being that is material, is caused, not by things outside the body, but by things within it, which operate outwards from within; that is to say, either by soul or by something else that is incorporeal... *--Ascl.* But surely, Trismegistus, it must be in void that things are moved. *–Herm.* You ought not to say that, Asclepius. Nothing that it, is void....[88]

To say that "the movement of the Kosmos" is caused not by things within its "body" or matter, but by "things within it", i.e., its "soul" or "incorporeal part" is tantamount to saying that space itself causes motion, since it is both (1) incorporeal, and (2) not a void. One may say that, for the ancient paleophysics, motion is the result of the interior potential of space. This interpretation is confirmed by the *Libellus II: 12a:*

> *Herm.* Now what was it that we said of that Space in which the universe is moved? We said, Asclepius, that it is incorporeal. *–Ascl.* What then is that incorporeal thing? *--Herm.* It is Mind, entire and wholly self-encompassing, free from the erratic movement of things corporeal.... The light whereby soul is illuminated.[89]

If one interprets this passage along the standard parameters of religious and occult interpretive techniques, then a contradiction emerges that is not resolvable:

---

[88] *Libellus II: 8b, 10,* Scott, op. cit., p. 139.
[89] *Libellus II: 12a,* Ibid., p. 141.

(1) on the one hand, "space" was described as not knowing itself, the primary distinction between it and God; and yet,

(2) on the other hand, "space" is now described as "Mind" which presumably is capable of knowing itself.

It is this contradiction that is the clearest indicator that one is *not dealing with religious or occult themes at all in the Hermetica of Trismegistus. One is dealing with physics.* The contradiction disappears if "mind" is interpreted in the sense of "all of the potential information in the field" of space. This subtle distinction between mind and space, and this interpretation of the passage, is confirmed by the *Libellus IV: 1b:* "the incorporeal is not a thing perceptible by touch or sight; it cannot be measured; it is not extended in space."[90]

The implication for physics is profound, for it means that in contrast to post-Newtonian or even Post-Einsteinian physical mechanics, the ancient paleophysics did not view the primary component of space as being extension or dimensionality, but rather as being *information.*

(11)
Magnitude:
Gauges of Order and Disorder:
Libellus VIII:3-4:

The *Libellus VIII:3-4* gives insight into yet another putative principle of the ancient paleophysics:

> Moreover, the Father implanted within this (sub-lunary) sphere the qualities of all kinds of living creatures.... For he wished to embellish with all manner of qualities the matter which existed beside him, but was hitherto devoid of qualities.... For when matter was not yet formed into body, my son, it was in disorder, which besets the small living creatures; for the process of growth and decay is a remnant of disorder.

---

[90] *Libellus IV: 1b*, Ibid., p. 149.

> The bodies of the celestial gods[91] keep without change that order which has been assigned to them by the Father in the beginning.[92]

One ventures down a much more speculative path with this passage, for here the remnant of the paleophysical principle is overgrown with a fecund and weedy growth of religion and metaphysics. The principle, however, may still be abstracted from the text:

(1) the Kosmos is a "body", i.e., a complex system of motion composed of several sub-systems of motion, and reacts to local disturbances much as a living organism;

(2) motion is change, a temporal process of "growth" from chaos to the order implied by a "system", and of decay to disorder from that system;

(3) the larger the system, the "larger" the time gauge of growth and decay for that system;

"Orders of magnitude" are thus orders of

(1) complexity
(2) time, and
(3) physical scale, or "size".

<u>*(11)*</u>
The "Seminal Reasons" (λογοι σπερματικοι, or rationes
seminales): Information in the Field:
Libellus: IX:6

The "patriarchal" and "masculine" imagery used by these texts for God is not arbitrary; it is a crucial metaphor, essential to the proper decryption of the ancient paleophysics, whether or not that metaphor was original to the paleoancient Very High Civilization,

---

[91] I.e., the planets and stars.
[92] *Libellus VIII: 3-4*, Scott, op. cit., p. 177.

or the result of the legacy civilizations trying to grapple with advanced conceptions. That patriarchal metaphor contains, in its overtly biological and sexual imagery, the power to convey specific information to the society with enough scientific and technological sophistication to understand it.[93]

> The Kosmos is an instrument of God's will; and it was made by him to this end, that, having received from God the seeds of all things that belong to it, and keeping these seeds within itself, it might bring all things into actual existence. The Kosmos produces life in all things by its movement....[94]

The theme of "seminal reasons" or seeds of reason is a persistent one in ancient Greek and Hellenistic philosophy, finding further extension and revision in the hands of ancient Christian writers such as Clement of Alexandria, Origen, and Athenagoras, and in the works of Church Fathers such as St. Justin Martyr, St. Basil of Caesarea, St. Augustine of Hippo, and St. Maximus the Confessor.

However, with the advent of genetic science and technology, the meaning of the metaphor achieves perhaps its final, or at least penultimate, clarity. The essence of the metaphor, if interpreted from the vantage of this scientific and technological plateau, consists of the implanting of specific information, "seeds", in the field, "Kosmos", highlighting yet again the profound connection between biology and physics within the ancient paleophysics.

---

[93] For other reasons, it is my opinion that the imagery was original to the posited paleoancient Very High Civilization and not to the subsequent legacy civilizations. It is perhaps significant in this regard that so much biblical injunction against idolatry was precisely directed at those practices that took the sexual and biological image in a literal sense that gave rise to the idolatry of fertility cults and so on. The injunction, in addition to its purely spiritual purpose, may have also had a mundane one, i.e., to learn to interpret such imagery properly when the right socio-scientific conditions prevailed for doing so, a difficult proposition if one is sacrificing babies and virgins to appease the bloodlust of the gods and ensure a good harvest! Of course, such a position more or less posits that the Biblical God was more or less the God of the putative ancient Very High Civilization. Making that case would be a complex and ambitious undertaking far beyond the purview of this work.

[94] *Libellus IX:6*, Scott, op. cit., p. 183.

<u>*(13)*</u>
<u>*The Soul and Instantaneous "Travel":*</u>
<u>*"Spooky Action at a Distance" and The Einstein-Podolsky-Rosen*</u>
<u>*Effect:*</u>
<u>*Libellus XI (ii): 19:*</u>

The final passage of the *Hermetica* with which one investigating any putative paleophysics must contend is the *Libellus XI(ii): 19:* "Bid your soul to travel to any land you choose, and sooner than you can bid it go, it will be there.... It has not moved as one moves from place to place, but it *is* there."[95] The "soul" as a manifestation of "mind" is realized information in the field. Something like an Einstein-Podolsky-Rosen effect is implied to be connected to this information.[96]

(F) David Hatcher Childress and Ancient Snaskrit "Vimana"
Texts:

Author, world traveler, and researcher David Hatcher Childress is yet another who maintains that a paleoancient Very High Civilization, possessed of the most advanced scientific and technological capabilities, once existed. Its achievements have been preserved, albeit in garbled form, in ancient texts and traditions, and in monuments and archaeological data, in this case in the texts and peculiar archaeological evidences left by the legacy civilization of the Indus Valley and ancient India. For Childress, however, the paleoancient Very High Civilization is his primary interest and focus, not the technology nor its putative scientific principles as such, much less an attempt to extrapolate them with reference to any particular site such as Giza. Nonetheless, a brief

---

[95] *Libellus XI(ii): 19,* Ibid., p. 221.

[96] An entirely different connection is possible to the techniques of "coordinate remote viewing" that was researched both in the United States and the Soviet Union during the Cold War. Of course, there may be, if one entertains a high degree of speculation, a connection between the EPR effect, Bell's non-locality theorem, and such "coordinate remote viewing".

review of his hypothesis, advanced in his book *Vimana Aircraft of Ancient India and Atalantis,*[97] is in order before turning to an examination of the Sanskrit text that comprises much of that book. Childress argues that these Sanskrit texts refer to an "ancient wisdom" and the vastly destructive purposes to which it was turned. Moreover, those destructive uses, he maintains, are corroborated by anomalous archaeological sites in the sub-continent that have all the earmarks of the use of nuclear and other weapons of mass destruction.[98]

Childress recounts various traditions of the sub-continent and the Far East that refer to whole libraries containing this "ancient wisdom":

> According to the famous astronomer Carl Sagan, a book entitled *The True History of Mankind over the Last 100,000 Years* once existed and was housed in the great library in Alexandria, Egypt.[99]

Moreover, "all ancient Chinese texts, especially those of Lao Tzu and Confucius (sic.) as well as the *I Ching*, speak of the ancients and the glory of their civilization."[100] Unfortunately, much of this knowledge was lost when the Emperor Chi Huang Ti ordered the destruction of all books relating to ancient China shortly before his death in 212 B.C.[101]

Further south, between China and the Indian subcontinent, similar traditions can be found pertaining to the Dalai Lama's palace, the Potala: "The prolific occult writer T. Lobsnag Rampa tells... of these underground tunnels beneath the Potala in his fascinating books *The Third Eye* and *Cave of the Ancients*." While the tunnels may have contained a library of ancient texts, Childress observes that Rampa's story is "somewhat dubious," serving only

---

[97] David Hatcher Childress, *Vimana Aircraft of Ancient India and Atlantis*(Adventures Unlimited Press, 1999), ISBN 0-932813-12-7.

[98] A similar proposition, it will be recalled, was maintained by Zechariah Sitchin in his *Wars of Gods and Men.*

[99] Childress, op. cit., p. 19. I have been unable to find the source of Sagan's remark.

[100] Ibid.

[101] Ibid.

to indicate the persistence of the myth surrounding such tunnels and hidden libraries.[102]

The China-Tibet-India connection is further underscored by the relation of a curious incident:

> Recently, Sanskrit documents discovered by the Chinese in Lhasa were sent to India to be studied by experts there. Dr. Ruth Reyna of the University of Chandigarh said that the manuscripts contain directions for building interplanetary spaceships!
>
> In any case, Dr. Reyna explained that the document stated that the method of propulsion was 'anti-gravitational.'...
>
> Indian scientists were at first extremely reserved about the value of these documents, but became less so when the Chinese announced that certain parts of the data were being studied for inclusion in their space program![103]

More to the point is his recounting of the famous incident when Alexander the Great's invasion of Indian Emperor Asoka's empire failed:

> It is also interesting to note here that Asoka's Empire in India was immediately after the attempted invasion of the Indian subcontinent by Alexander the Great, whose army retreated and all attempts were given up to subdue India, after his army was 'attacked' by what Greek historians later called, 'fiery flying shields.'[104]

The well-known Hindu epic, the *Ramayana* (literally, "Rama's Way"), in passages "thousands of years old" recounts wars with flying machines and weaponry akin to "particle beam weapons and horrifying explosive devices."[105]

---

[102] Ibid., pp. 24-25.

[103] Ibid., citing Robert Charroux, *The Gods Unknown* (New York City: Berkeley Books, 1969), no page reference given.

[104] Ibid., p. 27, citing Frank Edwards, *Stranger than Science* (New York City: Lyle Stuart, 1959), no page reference; and W. Raymond Drake, *Gods and Spacemen in the Ancient East* (London: Sphere Books, 1968), also no page reference.

[105] Ibid., p. 53.

In a few verses from the *Ramayana* and its companion epic, the *Mahabharata*, these weapons are described in all their destructive fury:

Gurkha, flying a swift and powerful vimana,
Hurled a single projectile,
Charged with all the power of the Universe,
An incandescent column of smoke and flame,
As bright as ten thousand suns,
Rose with all its splendour.

It was an unknown weapon,[106]
An iron thunderbolt,
A gigantic messenger of death,
Which reduced to ashes
The entire race of the Vrishnis and the Andhakas.

The corpses were so burned
As to be unrecognizable.
Hair and nails fell out;
Pottery broke without apparent cause,
And the birds turned white...."
                    The Mahabharata[107]

(It was a weapon) so powerful
That it could destroy the earth in an instant -
A great soaring sound in smoke and flames –
And on it sits death...
                    The Ramayana[108]

Dense arrows of flame,
Like a great shower,
Issued forth upon creation,
Encompassing the enemy....
A thick gloom swiftly settled upon the Pandava hosts.
All the points of the compass were lost in darkness.
Fierce winds began to blow.
Clouds roared upward
Showering dust and gravel

---

[106] Unknown to whom?

[107] Childress, op. cit., pp. 61-62, emphasis added.

[108] Ibid., p. 62, emphasis added.

> ...
> The earth shook,
> Scorched by the terrible violent heat of this weapon.
>
> ...
> *From all points of the compass*
> The arrows of flame rained continuously and fiercely.
> *The Mahabharata*[109]

While these texts are more than suggestive of the effects and results of the use of nuclear weapons – right down to "nuclear winter" scenarios and the loss of hair and teeth – certain phrases, if taken literally, are suggestive of the use of another type of weaponry altogether. The most significant of these suggestive phrases is the insistent of the texts that one was dealing not with many nuclear bombs, but with *one single weapon.* Other phrases suggest the "scalar" and "harmonic interferometry" physics being suggested as the basis of the Giza weapons complex:

(1) "all the power of the universe," suggesting a weapon reliant upon the physics of entangled systems;
(2) "the earth shook", suggesting the oscillation of the earth itself was a principle of the weapon;
(3) "arrows of flame" that radiated from "all points of the compass" to converge and bath a target in "terrible violent heat", suggesting the "ether force" that Tesla claimed to have discovered in his later experiments with his impulse direct current experiments.[110]

In summary, the Hindu epics describe, with an accuracy of detail not possible to any legacy civilization *not* possessing them, the use and results, at the very minimum, of nuclear weaponry. But more importantly, they suggest and would appear to corroborate independently the existence of a weaponized physics even more

---

[109] Ibid., emphasis added.
[110] More will be said on this in "The Weapon Hypothesis" chapter. It should also be noted that the four parabolic reflecting faces of the Great Pyramid are precisely aligned on the four cardinal compass points.

destructive than that, the physics of harmonically entangled systems and scalar interferometry.[111]

The strongest corroboration of these epics comes, however, not from the texts themselves, but from another quarter. While such passages are suggestive of a "paleoancient" Very High Civilization possessed of a science and technology advanced enough to construct and deploy such weapons of mass destruction, it is the archaeological evidence provided by the Indus Valley civilization that breathes archaeological and historical life into the epic poetry. Childress notes of these ancient cites that "archaeologists who have excavated the cities theorize from this that the cities were planned before they were built... even more remarkable is that the plumbing and sewage systems found throughout the 'Indus Valley Culture' are well laid out and planned."[112] This planning included private toilets, running water, and separate water and sewage systems, itself a standard not in evidence in the same region today.[113]

Then comes the bombshell:

> So the cities were sophisticated, but is there any evidence of the devastating wars spoken of in the Indian epics?... When archaeologists reached the street levels of these two cities during their excavation in the early fifties, they discovered skeletons scattered about the city, many just lying in the streets and some holding hands! It was as if some horrible doom had taken place, annihilating the inhabitants in one fell swoop. These skeletons are among the most radioactive ever found, on a par with those at Nagasaki and Hiroshima. At another site in India, Soviet scholars found a skeleton with a radioactivity level in excess of fifty times that which is normal.[114]

---

[111] Childress himself is alive to this possibility as he mentions "scalar wave weaponry" on pp. 88-89 of his work in conjunction with allegations in esoteric literature on the use of such weapons by "Atlantis". The best source of comments on scalar physics are the works of T.E. Bearden, Tesla Book Co.

[112] Childress, op. cit., p. 81.

[113] Ibid.

[114] Ibid., citing Richard Mooney, *Gods of air and Darkness* (New York City: Stein and Day, 1975); *The Atlas of Archaeology* (St. Martin's Press, 1982); Alastair Service, *Lost Worlds* (New York: Arco publishing, 1981); *Into the Unknown* (Reader;s Digest Assoc, 1981); Peter Kolosimo, *Not of This World*

Thousands of lumps, christened 'black stones', have been found at Mohenjo Daro. These are apparently, fragments of clay vessels that melted together in extreme heat and fused(sic.). Other cities have been found in northern India that indicate explosions of great magnitude. A city was found to have been subjected to intense heat. Huge masses of walls and the foundations of an ancient city were found fused together, literally vitrified![115]

The most puzzling archaeological datum concerning the apparent use of such weapons by the ancient Very High Civilization pertains to the approximate dating of such a civilization:

A news item that appeared in the New York Herald Tribune on February 16, 1947 (and repeated by Ivan T. Sanderson in *Pursuit*, January, 1970) reported that '(archaeologists) have been digging in the ancient Euphrates Valley (Iraq) and have uncovered a layer of agrarian culture 8000 years old, and a layer of herdmen culture much older, and a still older cavemen culture. Recently, they reached another layer of fused green glass.'[116]

When the first atomic bomb went off at Alamagordo in new Mexico, it turned the desert sand to green glass! Interestingly, Dr. Oppenheimer, the 'father of the H-Bomb,'[117] was also a Sanskrit scholar. Once when speaking of the first atomic test, he quoted the *Mahabharata* saying, 'I have unleashed the power of the Universe; now I have become the destroyer of worlds.' Asked at an interview at Rochester University seven years after the Alamagordo nuclear test whether that was the first atomic bomb to ever be detonated (sic.), his reply was: 'Well, yes,' and added quickly, 'in modern history.'[118]

Zechariah Sitchin cites similar archaeological evidence of the "paleoancient" use of atomic weapons.

(Seacaucus, New Jersey: University Books, 1971); no page references are given for any of these works.

[115] Ibid., pp. 81-82, citing all of the above sources but the last.

[116] Ibid., p. 82, citing Charles Fort, *The Book of the Damned* (Ace Books, 1919), no page reference given.

[117] Apparently a misprint, as Dr. Edward Teller is the "father of the H-Bomb." Childress meant to say "father of the A-Bomb."

[118] Childress, op. cit., p. 82, citing Charles Berlitz, *Doomsday 1999* (Doubleday, 1981), no page reference.

Childress is aware of the fact that standard academic archaeology is quick to dismiss such sites as evidence of the ancient Very High Civilization's use of nuclear or scalar weaponry, preferring to "explain" such "desert glass" sites, such as exist in Egypt's western desert, as the results of a meteor impact. Unfortunately, there is no evidence of an impact *crater* at this or any other such site, nor is there any corroborating theoretical evidence.[119]

The existence of texts -- such as the *Hermetica* or the ancient "pyramid" texts – that possibly contain the remnants of an ancient paleophysics is hard to explain from any conventional disciplinary perspective. Even more problematical, however, are the existence of texts that purport to explain, in detail, the existence of aircraft, texts that form much of the content of Childress' book and its central argument. The full title of this text is *Maharashi Bharadwaaja's Vymaanika-Shaastra or Science of Aeronautics.*[120] The foreword of that reprinted text reads as follows:

> On 25-8-1952 the Mysore representative of the press Trust of India, Sri N.N. Sastry, sent up the following report which was publishing in all the leading dailies of India, and was taken up by Reuter and other World Press News Services:
> 'Mr. G.R. Josyer, director of the International Academy of Sanskrit Research in Mysore, in the course of an interview recently, showed some very ancient manuscripts which the Academy had collected. He claimed that the manuscripts were several thousands of years old, dealing.... in elaborate detail about food processing.... One manuscript dealt with aeronautics, construction of various types of aircraft for civil aviation and for warfare.'[121]

---

[119] Childress, op. cit., pp. 82-83. The absence of craters, however, is typical for nuclear bombs, since they typically are deployed to explode in the air, rather than on the ground, at some distance above the target.

[120] Childress, p. 83. Throughout references to this text, citations are from Childress' reprint. The original is reprinted by Childress complete with its original pagination, which differs from the pagination of Childress' main text. Page numbers thus refer to the page numbers of the reprinted text.

[121] Ibid., p. i.

While most of these texts are written in the style of pseudo-technical gibberish, or relate details of "aircraft" that are simply impractical, nevertheless, here and there, as in the *Hermetica*, the "gibberish" could be the apparent result of a less scientifically sophisticated legacy civilization trying to preserve and understand the achievements and physics of a more sophisticated antecedent.

So, while one encounters the following claim of an interplanetary travel capability – "And Vishwambhara says: 'Experts say that that which can fly through air from one country to another country, from one island to another island, and from one world to another world, is a 'Vimana'"[122] -- this is followed by a "plausible context" that suggests not only very unconventional vehicle "airframes" but a range of "stealth technology," and the possible atmospheric ionization effects that would result from the use of strong electromagnetic fields, all in a culture that had never even heard of Maxwell's equations:

> The pilot should have had training in maantrica and taantrica, and antaraalaka, goodha or *hidden*, drishya and adrishya, *or seen and unseen*, proksha and aparoksha, *contraction and expansion, changing shape*, look frightening, look pleasing, *become luminous or enveloped in darkness*... stun by thunderous din, jump, move zig-zag like serpent,...face all sides...paralyse... or exercise magnet pull.[123]

It gets worse. The ancient "legacy" civilization apparently preserved the very sophisticated knowledge of its precursor in its awareness of levels of the earth's atmosphere:

> Goodha: As explained in the 'Vaayutatava-Prkhana', by harnessing the powers, Yaasa, Viyaasa, Prayaasa *in the 8th atmospheric layer convering the earth, to attract the dark content of the solar ray, and use it to hide the vimana from the enemy.*[124]

---

[122] Ibid., p. 2.

[123] Ibid., p. 3, emphasis added. It is not being suggested that electromagnetism is being used to propel such a craft, merely that electromagnetic effects are being observed in and perhaps manipulated by its operation.

[124] Ibid., p. 3.

What is amazing is the apparent accuracy of this knowledge. The standard atmospheric layers are the troposphere, stratosphere, mesosphere, thermosphere, and exosphere. If one includes the ozone layer between the stratosphere and mesosphere, that gives six "layers". The clue to the two "missing" layers is provided by the passage itself, since it suggest the manipulation of solar radiation in conjunction with electromagnetic phenomena. The Van Allen radiation belts, an inner belt approximately 1000 to 5000 kilometers about the equator, and an outer belt at approximately 15,000 to 25,000 kilometers, consist of protons and electrons captured by the earth's magnetic field from the solar wind or produced as products of cosmic ray bombardment. Once captured, these particles oscillate between the magnetic poles, spiraling around the field lines and emitting radiation as they do.[125]

Other texts suggest deliberate knowledge of the relationships of the solar wind, the magnetosphere, and the electro-dynamics of plasma phenomena such as the Aurora Borealis:

> Drishya: by collision of the electric power and wind power in the atmosphere, a glow is created, whose reflection is to be caught in the vishwa-kriyaa-drapana or mirror at the front of the vimana, and by its manipulation produce a maaya-vimana or camouflaged vimana.[126]

It is important to recall that similar plasma phenomena were observed by Tesla in his later experiments, as well as alleged in connection to the Philadelphia experiment.[127]

---

[125] The text suggests that a component of 'stealth technology" is thus precise knowledge and manipulation of magnetospheric and Van Allen-solar wind phenomena, i.e., suggests a very precise knowledge of the magnetic field lines of each point in the earth's surface is required in order for stealth technology to work.

[126] Ibid., p. 4.

[127] The principle of magnetic image resonance might be modified or configured in such a fashion to duplicate or manipulate such phenomena. The *DE Eldridge* was said to have been surrounded by a green glow or plasma, an indicator of extremely strong electromagnetic ionization of the atmosphere, prior to its alleged disappearance and reappearance.

Other texts suggest the physics of "harmonic systems entanglement" of locally isolated systems, such as was encountered in the *Hermetica:*

> Adrishya: According to 'Shaktitantra', by means of the Vynarathya Vikarana and other powers in the heart centre of the solar mass, attract the force of the ethereal flow in the sky, and mingle it with the balaahaa-vikarana shakti in the aerial globe, producing thereby a white cover, which will make the vimana invisible.[128]

As the two previously cited texts also suggest, there was also apparently enough sophistication in the ancient paleophysics and its "paleo-stealth technology" to distinguish between practical applications that resulted in mere "camouflage" and in outright "invisibility."

Yet other texts suggest either the destructive use of electromagnetic pulse-type weapons, or some sort of electromagnetic rail guns:

> Pralaya: As described in the magic book of destruction, attract the 5 kinds of smoke through the tube of the contracting machine in the front part of the vimana, and merge it in the cloud-smoke mentioned in 'Shadgarbha Viveka', and pushing it by electric force through the five-limbed aerial tube, destroy everything as in a cataclysm.[129]

Not only does this passage suggest electromagnetic weaponry, it is also extremely suggestive of the jet engine, and these seemingly unrelated points require some digression.

Some have speculated that the B-2 stealth strategic bomber employs more than just the flying-wing airframe and advanced RAM (radar absorbent materials) technology, but that its most classified elements concern its propulsion system, which is more than a merely conventional jet engine. In a work entitled *Electrogravitics Systems: Reports on a New Propulsion Methodology*, by Thomas Valone, M.A., P.E., an article by Paul A

---

[128] Childress, op. cit., p. 4.
[129] Ibid., p. 5.

LaViolette, called "The U.S. Antigravity Squadron" is reproduced in full. It is worth citing the abstract of the article *in toto:*

> Electrogravitic (antigravity) technology, under development in U.S. Air Force black R&D programs since late 1954, may now have been put to practical use in the B-2 advanced technology bomber to provide an exotic auxiliary mode of propulsion. *This inference is based on the recent disclosure that the B-2 charges both its wing leading edge and jet exhaust stream to a high voltage. Positive ions emitted from its wing leading edge would produce a positively charged parabolic ion sheath ahead of the craft while negative ions injected into its exhaust stream would set up a trailing negative space charge with a potential difference in excess of 15 million volts. According to electrogravitic research carried out by physicist T. Townsend Brown, such a differential space charge would induce a reactionless force on the aircraft in the direction of the positive pole. An electrogravitic drive of this sort could allow the B-2 to function with over-unity propulsion efficiency when cruising at supersonic velocities.*[130]

The similarity of LaViolette's article abstract to the Sanskrit text is both suggestive and provocative, not the least because it suggests that one "source for suggestions" for such black projects is the "paleophysical" investigation of such texts.[131]

---

[130] Paul A LaViolette, "the U.S. Antigravity Squadron," cited in Thomas Valone *Electrogravitics Systems: reports on a New Propulsion Methodology* (Integrity Research Institute Publishers, 2nd edition, 1995), p. 82, emphasis added. The original source of the disclosure, as LaViolette notes in his article on pp. 82-83, was a group of "renegade" scientists and engineers involved in black R & D projects. The disclosure was made in the March 9, 1992 issue of *Aviation and Space Technology.* LaViolette also observes on p. 87 of the same article that "One indication that Brown's electrogravitics ideas were being researched by aerospace industry surfaced in January 1968. At an aerospace sciences meeting held in New York, Northrup officials reported that they were beginning wind-tunnel studies to research the aerodynamic effects of applying high-coltage charges to the leading edges of aircraft bodies," citing "Northrup studying sonic boom remedy," *Aviation Week and Space Technology,* Jan 22, 1968, p. 21.

[131] Other passages of the Sanskrit text as suggest coupling of jet engines and electrostatics are to be found on pp. 6, 38, and 93 of Childress' work. The more disturbing implication of such technology and its possible investigation and application by the Third Reich are explored in Renato Vesco & David Hatcher

The similarities accumulate. One Sanskrit passage, while seemingly gibberish, may provide important insights into the state of materials engineering in the "paleo-ancient" Very High Civilization:

> Zinc, sharkara or quartz poweder?...vebra or red-lead, yellow thistle...should be powdered...filled in a shashmooka crucible, placed in mandooka furnace, and with five-mouthed bellows heated to 200 degrees and melted to eye-level, when cast will yield a fine, light, blue bydaala alloy.[132]

The text is suggestive because its hints at a technology sophisticated enough to know of powdered states of certain metals plus a technique called "sinterization."[133]

LaViolette notes that one aspect of the B-2's highly classified technology lies precisely in the engineering not only of its enhanced propulsion efficiency, but in its radar invisibility, precisely the same sort of technology implied by the Sanskrit text:

> Authorities tell us that the hull (of the B-2) is composed of a highly-classified radar-absorbing material (RAM)....
>
> Evidence that the B-2 might indeed use a high-density ceramic RAM comes from information leaked by the above mentioned black world scientists who disclosed about the development of low-radar-observability dielectric ceramics made from powdered depleted uranium. The material is said to have approximately 92% the bulk density of uranium, which would give it a specific gravity of about 17.5, as opposed to 6 for barium titanate dielectrics. Thus, this new material has about three times the density of the high-k ceramics... and hence would develop at least three times the electrogravitic pull.[134]

Childress, *Man-Made UFOs, 1944-1994: Fifty Years of Suppression* (Kempton, Illinois: Adventures Unlimited Press), 1994.

[132] Childress, op. cit., pp. 72-73.

[133] Sinterization is a comparatively recent technology in metallurgy. Explored by the Allies and the Nazis prior to World War Two, the process consists mainly in producing metal alloys with micro-porous surfaces. The Germans, in particular, investigated the use of such surfaces in order to suction the boundary layer of air lift surfaces in order to increase the lift of airframes.

[134] LaViolette, "U.S. Antigravity Squadron," p. 92.

That is, the depleted uranium ceramic dielectric hull of the B-2 is the indicator of an emerging "unified technology" resulting from a different approach to the physics of the relationships of fluid dynamics, electromagnetism, and gravity! One the one hand, its dielectric properties render it invisible to radar, on the other, the build up of an electrostatic potential between the wing's leading edge and the jet exhaust increases the difference of charge, increasing the efficiency of the jet exhaust by inducing a "gravity well" at the front of the aircraft and a "gravity hill" behind it, which further reduces drag and turbulence, which further enhances radar invisibility, and so on and so on.

Returning now to the Sanskrit, there is further evidence of the sort of "harmonic systems entanglement" physics believed to be embodied at Giza:

> The gravity of the centre of the earth, the gravity of global earth, the solar flood, the air force, the force emanating from the planets and stars, the sun's and moon's gravitational forces, and the gravitational force of the universe, all together enter the layers of the earth in the proportion of 3,8,11,5,2,6,4,9 and, aided by the heat and moisture therein, cause the origin of metals, of various varieties, grades and qualities.[135]

This passage is rich in possible paleo-physical allusions, so it is best to summarize them in terms of a table of possible meanings:

Harmonic Systems Entanglement:
A Table of Possible Meanings of a Sanskrit Passage:

Phrase	Possible Physical Principle:
1. "the gravity of the center of the center of the earth"	1. The terrestrial (base planetary system's) theoretical center of gravity
2. "the gravity of global earth"	2. Actual gravitational

---

[135] Childress, op. cit., p. 16.

3. "the solar flood"

4. "the air force"

5a. "the force emanating from the planets"

5b. "...and stars"

6. "and the gravitational force of the Universe

7. "All together enter...the earth in the proportion of 3,8,11,5,2, 6,4,9..."

8. "and, aided by the heat and moisture therein, cause the origin of metals of various varieties, grades, and

acceleration at various points on the terrestrial (base planetary system's) surface

3. The electromagnetic "solar wind" of the base solar system as captured by the base planetary system

4. Several possibilities:
   a. The "aether" or ZPE
   b. The natural atmospheric condensor of the earth
   c. The dielectric "cavity" between the earth and the ionopshere
   d. Any combination of the above

5a. The inertial mechanical energy and harmonics of base solar system

5b. The inertial mechanical energy and harmonics of the base galactic system

6. The Total System with all its entangled Sub-Systems

7. The means of entanglement of non-local systems is "proportional" i.e., harmonic

8. All matter arises from sub-quantum fluctuations of the ZPE, given certain conditions of the thermal

qualities"   and temporal gradient.

In addition, the properties of this "systems entanglement physics" are specifically vorticular, and moreover, a basic phenomenon of the macrocosm and microcosm:

> Aavartas or aerial whirlpools are innumerable in the above regions. Of them the whirlpools in the routes of the vimanas are give.... Whirlpool of energy...whirlpool of winds... whirlpool from solar rays....whirlpool of cold currents.[136]

The fact that electromagnetic plasmas only a few millimeters in diameter can form patterns remarkably similar to the patterns of spiral galaxies millions of light years across suggests that there is a scale invariance of vorticular phenomena from the level of quantum particle collisions to the formations of whole clusters of galaxies into similar spiral filaments. This scale invariance is basic to the "plasma cosmology" emerging in the work of Swedish physicist Hannes Alfvén.[137] The importance of this vorticular phenomenon across degrees of scale from very large to very small is of crucial importance, for it suggests an avenue toward the unification of the various fields of physics, an avenue not being adequately explored.

Another text indicates that some sort of "unified field physics" not only once existed, but that its basis was the "as above, so below" harmonic entanglement of systems encountered at Giza and in the *Hermetica*:

> A foot-plate; 23 main centres to be marked on it, with lines connecting the centres. Similar number of revolving screws, wired tubes, pole with three wheels, eight liquids, eight crystals, eight liquid containers, mirror to attract the forces of shireesha, cloud, earth, stars, and aakaasha...[138]

---

[136] Childress, op. cit., p. 8.

[137] Cf. Eric J. Lerner, *The Big Bang Never Happened* (New York: Vintage, 1992), pp. 15-47, especially pp. 39-47.

[138] Childress, op. cit., p. 36.

Moreover, the "crystals" are to be immersed in acids to produce what may best be described as a "harmonically modulated" condenser.[139]

More suggestive is a passage that implies that the forces of the base solar system can be unlocked destructively on earth:

> During the passage of sun and other planets in the 12 houses of the Zodiac, owing to the varying speeds of their progressive and retrogressive motions, conflicting forces are generated in the Zodiacal regions, and their collisions will let loose floods of fierce forces which will reduce to ashes the parts of the plane which get involved with them.[140]

That is, one implicit assumption of "harmonic entanglement physics" is that the destructive release or use of such forces can only occur at certain harmonically dictated conditions, i.e., at certain times.

Other vorticular phenomena implied by the Sanskrit texts are superconductivity,[141] and the so-called and much-maligned "mercury vortex engines" said to be used in vimana aircraft.[142] Something like the plasma physics and cosmology of Alfvén seems to be definite component of the ancient paleophysics. Eric J. Lerner relates this episode from Alfvén's years in Sweden:

> In the late fififties, Alfvén and others... had been called in by the Swedish power company, ASEA, to solve an urgent problem. Most of Sweden's electrical supply is generated by hydroelectric power in the north of the country, then transmitted over six hundred miles to the industrial south. ASEA found that it was cheaper to transform the alternating current to direct current for transmission with large mercury rectifiers. A rectifier allows a current to pass in only one direction, holding it back for the other half of the cycle, thus producing DC. But every so often a rectifier would explode....

---

[139] Ibid., p. 37.
[140] Ibid., p. 41.
[141] Ibid., pp. 49-50.
[142] Ibid., p. 251.

Herlofson and Alfvén were consulted because the rectified mechanism, consisting of a low pressure mercury vapor cell, employs a current carrying plasma. The team from the Royal Institute rapidly located the problem: the pressure of the mercury vapor in the rectifiers was too low. As a result, at high currents nearly all the electrons carried the electrical flow, creating an unstable situation in which the plasma started to slosh about within the rectifier.

At low current, this sloshing was not serious. If too many ions (the positive charges) piled up on one side, the electrons would be attracted to them, neutralizing them. But at high current something else happened. If the ions accidentally spilled out of a region, the electrons would rush at them with such momentum that their collision pushed the ions farther out of the region. This accelerated the electrons more, and so on. However, a few ions would break away and accelerate toward the electrons, pushing most of them back. An ever-widening tear in the plasma would open up, with electrons bunching up on one side and ions on the other. As the gap widened, fewer electrons could pass, so the current... would drop. This is like suddenly unplugging an appliance. The drop in current produces a sudden drop in the magnetic field created by the current, and the changing magnetic field creates a powerful electrical force that further accelerates the electrons. In the case of an unplugged appliance, the voltage becomes high enough to make a spark jump from the socket to the plug. In the case of the rectifier, the voltage builds and builds until the electrons heat the rectifier plasma so hot that an explosion ensues, and gigantic sparks jump through the air in the station.[143]

In other words, in terms of the systems analysis and entanglement being proposed to understand the paleophysics, the rectifier becomes an over-unity device curling energy back into a system that becomes *closed* (i.e., non-entangled), compelling more energy to be curled back into the system, until a threshold is crossed and it self-destructs.

---

[143] Lerner, op. cit., p. 197. One cannot help but be taken with Lerner's description of the problem confronting the Swedish physicists and its similarities to the later high voltage "impulse" experiments of Nicola Tesla.

Some Conclusions

To the physicist, the foregoing must seem at the very least perplexing. So much of the putative principles of the paleophysics seems so far-fetched, and yet, so much of it very familiar to the concepts and models to be found within physics within the last forty years or so.

And to the layman, the technical terms and concepts alluded to must seem forbidding at the very least, and perhaps incomprehensible. But in either case, I hope sufficient material has been presented to persuade both the expert and the layman that there did indeed once exist an extremely sophisticated Civilization with a commensurately advanced physics and engineering.

The texts we have examined lead us to the following conclusions:

- That the Pyramid was a weapon of mass destruction (Sitchin);
- That its destructive power exceeded nuclear weaponry (Sitchin);
- That the Giza compound was a "royal fortress," i.e., a military complex (Hancock);
- That its most crucial components are missing or destroyed (Sitchin);
- That there once existed an extremely sophisticated physics (Plato, the *Hermetica*, Sanskrit "vimana" texts); and,
- That the sophisticated paleophysics was weaponized in weapons of mass destruction and that these weapons were used (the Hindu Epics).

But what was that paleophysics exactly? How can modern physics help to unlock its secrets? Those questions are addressed in the next chapter.

IV.
De Physica Esoterica

"Ironically, although superstring theory is supposed to provide a unified theory of the Universe, the theory itself often seems like a confused jumble of folklore, random rules of thumb, and intuitions."
Michio Kaku, Introduction to Superstrings and M-Theory[1]

"In theoretical physics, simply being brilliant is not enough. One must also be able to generate new ideas, some of them bizarre, which are essential to the process of scientific discovery."
Michio Kaku and Jennifer Thompson, Beyond Einstein: the Cosmic Quest for the Theory of the Universe[2]

The history of modern theoretical physics is breathtaking, and its technological impact on society is the visible sacrament of its progress. A little less than five hundred years ago when Sir Francis Bacon first proposed the outlines of scientific method, and when Sir Isaac Newton took the bold step of a mathematical model of a theory of gravity, men calculated with pencils to candlelight. A little less then one hundred years ago, analog computers on moving battleships calculated how to lob a one ton artillery shell from a moving platform to a moving target the size of a football field over twenty miles away. A little less than sixty years ago, the precursors of the first digital computers cracked the "uncrackable" permutations of the German enigma code and calculated how much critical mass was needed to split the "unsplitable" atom. Their descendants calculated how to lob nuclear and thermonuclear bombs thousands of miles. A little more than forty years ago, man learned how to defy the inverse square law and concentrate mere light into a beam that can cut through steal. More recently, we

---

[1] Michio Kaku, *Introduction to Superstrings and M-Theory,* Second Edition (New York: Springer-Verlag, Inc.: 1999), p. vii.
[2] Michio Kaku and Jennifer Thompson, *Beyond Einstein: The Cosmic Quest for the Theory of the Universe* (New York: Anchor Books, 1995), p. 65.

have learned how to entangle photons and send information from one location to another at faster-than-light speeds.

Scientists and medical professionals once thought man could never withstand the force of travelling faster than 30 miles per hour. Then came the railroad and man traveled 60 miles an hour on a regular basis. Physicists and engineers said man would never fly. Man flew. They said that man would never fly faster than sound. Man now flies at speeds well above the speed of sound. Scientists said we'd never reach the moon, and we did. If anything emerges from this, it is that science is ever-changing, and that change is driven by the all too human desire to find some way to accomplish what some say cannot be accomplished. At one time, science said that everything was harmonically related to everything else. The whole Universe was a dance to the "cosmic harmony of the spheres." That doctrine is now held by mainstream science to be naïve "Pythagorean" notion of a by-gone age of Greek metaphysics.

And a little more than one hundred years ago, science believed in a super-fine matter called "aether lumeniferous", literally "light bearing stuff", as the carrier medium for light and other electromagnetic waves. No less a scientist than James Clerk Maxwell believed in the idea. That notion, too, was exploded in a famous experiment by two American physicists named Michelson and Morley, and an even more famous theory by a German-Jewish scientist working as a clerk in a Swiss patent office named Einstein. Out of that paradigm-shattering experiment and the equally revolutionary theory that followed were erected two of the stable paradigms of modern physics: there is no aether, and nothing can travel faster than the velocity of light.

So let us examine each of these three things – the "cosmic harmony" or "symphony," the aether lumeniferous, and the shattering experiment and theory – as a way of entering the discussion of the more esoteric notions of contemporary experimental and theoretical physics.

A. The Cosmic Harmony:
The Harmonic Series and Coupled Oscillators

Anyone remotely familiar with music, and particularly keyboard music, already knows about the harmonic, or overtone, series. Even though they may not know the *term*, they know the *phenomenon*. It may be easily demonstrated by sitting at an acoustic piano.[3] Imagine you are sitting at a piano. Now, press down the note middle c *silently*, and hold the key down, allowing the string inside the sound box to vibrate. Then, strike another note c, an octave or two lower. What you will hear is the note you are pressing down silently vibrating sympathetically with the note you struck.

The reason for this is simple. The length of the piano string of the *struck* note vibrates not only in waves the exact length of the string, but in *fractions* of that length, in halves, thirds, fourths, fifths, and so on. To see that this is true, repeat the experiment, only *this* time, press down the note g and hold it silently, and again strike the note c somewhere below it. Again you will hear the note g, a bit fainter than the silently pressed c earlier, but you will hear it none the less. In fact, you can press *any* note silently, and strike any *other* note, and you will hear the silently pressed note. All the silently pressed notes on the keyboard are thus said to be *overtones* or *harmonics* of the struck note, which is the *fundamental*. When they vibrate along with the fundamental, they are said to be sympathetic or resonant with the fundamental. When we press the note silently, we have created what physicists call a "coupled harmonic oscillator," because the silently pressed note oscillates to the note that is struck; it is coupled with it. Put in this way, a keyboard instrument such as a piano or a pipe organ is nothing but a collection of coupled harmonic oscillators designed to harness the vibratory energy of strings or columns of air in order to produce the regular waves that we call musical tones.

---

[3] An acoustic piano, rather than a digital or electric one, is necessary in order to demonstrate the overtone series.

113

Once the ancients had discovered this principle, and the mathematical properties governing it were understood, it was a natural conclusion to speculate that everything that was in motion was somehow coupled with everything else in motion. In short, the universe itself was but a collection of coupled harmonic oscillators. So what went wrong?

Very simply, sound could not travel where there was no medium of propagation, whether it is air, strings, stretched animal hides, or whatever. And once vacuum space was discovered, the "cosmic harmony" as a way of looking at the universe in a unified fashion was on the way out. The paradigm had shifted.

B. *The Aether Lumeniferous*

Old ideas have a strange was of coming back to life, and the notion of the cosmic harmony was no different. Once light and other electromagnetic phenomena were discovered to travel in *waves* just like the acoustic phenomena of sound, science appeared to be confronted with a dilemma. Every wave phenomenon known to science appeared to require some medium of propagation, some form of matter to which the wave could "attach" or on which the wave could "ride." But science knew that light traveled through space, which was a vacuum, and apparently devoid of matter such as we know it. Thus came the apparently rational conclusion: there had to be some form of matter, a superfine matter or "stuff", existing in the vacuum as well as within the "pores" of matter such as glass or water, that acted as the propagation medium for light and other electromagnetic waves. This was called, appropriately enough, the "aether lumeniferous", or "light-bearing stuff." The mistake is easy enough to see now: scientists were merely extending their understanding of acoustic wave phenomena to light itself. They were thinking of *light* as if it were *sound*.

On that basis, a rather logical chain of reasoning was erected. Imagine one is in a train, and that the train is traveling at x miles per hour. If one stood and then walked in the same direction as the train is traveling, i.e., toward the front of the train, at the speed n, then one would be traveling at x + n miles per hour. By the same token, if one walked toward the rear of the train, one would still be traveling in the same overall direction as the train, but at x – n miles per hour. This process is known to the physicist as "vector addition". A vector is used to measure velocity, and a word is necessary about this.

To the physicist, velocity is not the same thing as speed. Speed is simply "how fast", but velocity is not only "how fast" but also "which direction." Thus, when our walking passenger is traveling the same direction as the train, his velocity is added to the velocity of the train to obtain his true speed. But when walking in the opposite direction, his velocity is subtracted from the train's velocity to obtain his true speed. A "vector" is therefore, for our purposes, almost synonymous with velocity.

C. The Shattering Experiment and the Revolutionary Theory

To appreciate what all this means for the theory of an aether lumeniferous, we now have to go back to the fact that scientists were thinking of light like they were thinking of any other conventional system where vectors may be added or subtracted. Scientists reasoned that if they could split a beam of light, and send one part of the beam traveling in the same direction as the earth's rotation and the other part counter to it, they should be able to detect this aether by the simple process of vector addition.

Why this is so takes a little explaining. The aether itself was understood to be of such very fine structure that it remained stationary in space. Thus, when a beam of light was shot in the same direction as the earth's rotation, they reasoned that the aether would "blow" past the earth's surface. This blowing they called the "aether wind". As it blew, it would slow the beam traveling against it down, and speed up the beam traveling in the same direction. A measurable difference in the velocity of light between these two beams, however small, would verify the existence of the aether.

And so two American physicists designed what is probably the most famous experiment in twentieth century physics, the Michelson-Morley experiment. To understand it adequately, we must also recall another fact about pre-relativistic physics: scientists thought of light as if it were sound. While most people may not know the term, most are again familiar with the phenomenon known as the Doppler Effect.

Imagine you're standing beside a railroad track, and a fast-moving express train is approaching you, blowing its horn continuously. Since the vectors of the speed of the train and the speed of the sound waves of the horn are added, the sound waves themselves are more compressed, and the pitch of the sound is appropriately higher. At the instant the train passes you, the vectors are subtracted, the waves are stretched, and the pitch falls. The frequency of the sound waves – at least for the observer on the ground – shifts. This is called "phase shift."

We are now in a position to understand the Michelson-Morley experiment. Diagram one, from Stan Deyo's *The Cosmic Conspiracy*, is a schematic of the experiment. Deyo comments as follows:

> It was reasoned that by splitting a beam of light (F) into two parts; (sic.) sending one out and back in-line with the direction of the earth's orbital path, (to mirror A) from half-silvered mirror (G); send the other at right angles to the direction of earth's orbital path (to mirror B) through half-silvered mirror (G) and glass plate (D); and recombining the two beams in the interferometer (E) one should be able to detect a shift in the phases of the two beams relative to one another.[4]

Note how diagram one and the way the experiment was configured conform to the understanding of vector addition and the Doppler Effect. Think of the arrow to the right representing the "direction of ether wind" as the railroad locomotive traveling toward the interferometer, which is you. Now note that the beam of light-sound coming back from B is combined with the beam of light-

---

[4] Stan Deyo, *The Cosmic Conspiracy*, New Revised Edition (Kempton, Illinois: Adventures Unlimited Press, 1994), p. 169.

sound coming back from A. It is as if the experiment was designed to allow you to hear both the sound wave traveling toward you and away from you *at the same time.* The difference in "sounds" is the phase shift, allowing you to hear two different tones, or to see two slightly different "colors" or wave patterns. If you see the different patterns, a shift has occurred, and the ether drag is verified, and with it, the existence of the ether.

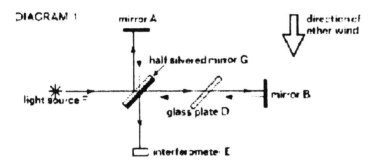

But note, however, that the experimental apparatus itself is likewise traveling along with the earth. So Diagrams 2a and 2b represent the actual path of the light beams.

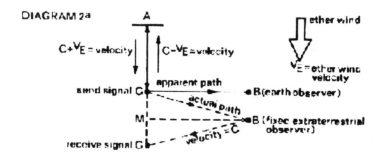

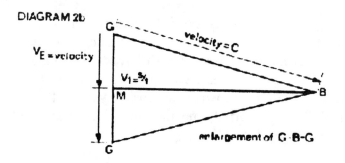

DIAGRAM 2b

V_E = velocity

velocity = C

$V_1 = \frac{s}{1}$

M

B

G

enlargement of G-B-G

The results of the experiment are now well known: no phase shift was detected, and therefore the velocity of light remained paradoxically the same in both directions. What followed, however, was a paradigm shift in theoretical physics, the ripple effects of which are found all the way down to current superstring theory, based as it is on the theoretical assumptions that Einstein made to interpret the results of the experiment.

To understand what happened, the mathematics are less important *than the assumptions that undergirded them.* The reader who is bewildered, or at the least, distressed by the mathematical presentation is advised to *realize that the interpretation of the equations, which itself is not mathematical but of a deeper philosophical nature, is where the real importance of the experiment and Einstein's theory lie.*[5]

When the results of the experiment violated the theoretical expectation dictated by the aether paradigm, two physicists, Lorentz and Fitzgerald designed a series of equations known as the Lorentz Transforms designed to bring the test results and the hypothetical expectation into agreement.[6] What these equations did, in effect, was to say "that length shortened, mass flattened, and time dilated as a body moved through the ether; hence it was possible to detect the ether (Sic.)" by these means.[7]

The transform equations of Lorentz made their way into Einstein's theory of Special Relativity. Einstein interpreted the

---

[5] For those inclined to want the mathematics, cf. Deyo, op. cit., pp. 171-174.
[6] Ibid., p. 172.
[7] Ibid.

data of the experiment *to men that the velocity of light was a uniform constant to any observer.* With that, the notion of an aether in the sense that had obtained up to that point was discarded for the simple reason it was no longer needed. What was retained in his theory, via the transform equations, were the time dilations and length contractions themselves, which were now interpreted to be the result of the acceleration of any mass to near light velocity. Moreover, the transforms were essential to Einstein's derivation of the now famous equation $E=Mc^2$. [8]

Deyo's commentary at this point is breathtaking:

> (Einstein's) failing was in declaring the velocity of light an observable limit to the velocity of any mass when it should only have been the limit to any observable electromagnetic wave velocity in the *ether* (sic.). The *velocity of light is only a limit velocity in the fluid of space where it is being observed.* If the energy-density of space is greater or less in another part of space, then the relativistic velocity of light will pass up and down through the *reference light wave velocity limit* – if such exists.
>
> ...When a fixed-density fluid is set in harmonic motion about a point or centre (sic.), the *number of masses passing a fixed reference point per unit time* can be observed as increased mass (or concentrated energy). Although the density (mass per volume) is constant, the mass-velocity product yields the illusion of more mass per volume per time.[9]

In other words, in the very act of using transforms that were themselves the product of the paradigm of belief in the aether, Einstein did not remove the notion; he *relied upon it* but forgot that he was doing so. The error lay not in the mathematics, but in his interpretive logic. As a result, the velocity of light was misinterpreted as a constant "upper boundary limit" on velocity for all masses, rather than as *a boundary condition between different types of fluid spaces coexisting in the same system.*

---

[8] Ibid., p. 172.
[9] Ibid., p. 174.

D. Back to Michelson-Morley: Measuring the Wrong Thing at the Wrong Place and Drawing the Wrong Conclusions

(1) Sagnac's Rotational Version of Michelson-Morley

The effects of relativity on the subsequent history of theoretical physics down to our own time have thus been profound and abiding. But like all such paradigm shifts, a few physicists refused to accept it, and their efforts and experiments were, initially at least, brushed aside. One of them, however, was a French physicist by the name of Georges Sagnac, who proposed that Michelson and Morley had measured the wrong thing at the wrong place. After all, the Michelson-Morley experiment dealt with split perpendicular beams traveling in straight linear paths. But, he reasoned space is not like that. Space, and the objects in it, *rotate.* It seemed logical to assume, then, that the aether rotated as well. Therefore, if one wished to detect it, one had to measure the velocity of a split beam of light in a *rotating* experiment (cf. Diagram 3).

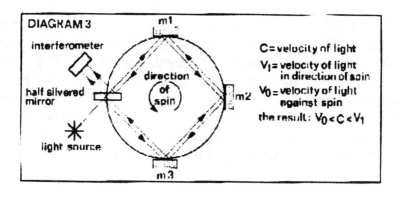

(2) Deyo's Teacup Analogy

In other words, there was a fundamental error in the very way the Michelson-Morley experiment was set up.

> The *error* of the M-M experiment is *the test results are also valid for the case where there is an ether* (sic.) *and it, too, is moving along with the same relative velocity and orbit as Earth maintains around the Sun.* The *tea cup* analogy can be used to explain the error. If one stirs a cup of tea (preferably white) which has some small tea leaves floating on its surface, one notices some of these tea leaves orbiting the vortex in the centre (sic., et passim) of the cup. The leaves closer to the centre travel faster than those farther from the centre (both in linear and angular velocity).
>
> Now, one must imagine himself greatly reduced in size and sitting upon one of these orbiting leaves. If one were to put his hands over the edge of his tealeaf on any side, would he feel any tea moving past? .... No. The reason is that the motion of the tea is the force that has caused the velocity of the leaf. One could not detect any motion if both he and the tea were travelling in the same direction and at the same velocity. However, if one had arms long enough to stick a hand in the tea closer to either the centre or the rim of the cup where the velocities were different to his own, then he would feel tea moving faster or slower than himself (respectively).
>
> Also, if one were to spin his tea leaf at the same time as it orbits about the centre, placing his hands into the tea immediately surrounding his leaf would show inertial resistance against the spin moment of his leaf.[10]

That, essentially, is the reasoning Sagnac used to set up his rotational version of the Michelson-Morley experiment. And the result was exactly what the aether paradigm predicted. Sagnac's experiment successfully demonstrated that "the velocity of light sent *in* the direction of spin around the perimeter of a spinning disc (for the surface of the earth) varied from the velocity of light sent *against* the spin."[11]

---

[10] Deyo, op. cit., pp. 176-177, emphasis in the original.
[11] Ibid., p. 176.

Sagnac's experiment was repeated by Gale and Michelson himself in 1925, and with the same results. And more recently, scientists have repeated it using a ring-laser system and the more precise modern measuring methods. And the result is always the same: *the velocity of light is **not** constant.*[12]

The effect of Sagnac's experiment on theoretical physics is difficult to assess. On the one hand, Einstein most certainly probably formulated General Relativity in part to address its results. This is evident in the conclusion that local space-time around rotating large masses is curved, and can in turn bend light. However, some of the mistaken assumptions drawn from the linear version of the Michelson-Morley experiment were *not* discarded, and their effects persist. For example,

> In the preceding analogy, the centre of the spinning tea (or vortex centre) represented the Sun, leaf: the earth; the tea: the ether; and the rider's hands: the light *beams* of the M-M test. In essence, what Michelson, Morley, Einstein and many other scientists have said is that the M-M test showed the velocity of light was not *affected* by the Earth's orbital motion. *"Therefore,"* they have said, *"we have one of two conclusions to draw":*
>
> *1)   The Earth **is orbiting the Sun** and there is no ether, or,*
>
> *2)   The Earth is not orbiting the Sun and there is an ether but since the earth is not moving through the ether, the ether "wind" cannot be detected.* Obviously, this conclusion is negated by Earth's observed helio-centric orbit.
>
> However, their reasoning should also have incorporated a third option:
>
> *3)   **The Earth is orbiting the Sun and so is the ether;** therefore, no ether wind could be detected in the orbital vector immediately in the vicinity of Earth.[13]

---

[12] Deyo, op. cit., p. 177. An interesting implication seems to be implied, one which I am not aware has been tested experimentally: does the velocity of light vary as some function of the speed and/or mass of the rotating system? General relativity would seem to imply that some such relationship is possible. Indeed, Einstein in part formulated general relativity as a means of addressing the issues raised by Sagnac's experiment.

[13] Ibid., p. 177.

Put differently, every large rotating mass drags, or whirls, the aether along with it and around it. This will become a crucial consideration when we consider Bell's non-locality theorem, for as we shall discover, *this presupposition of a vorticular structure to the aether itself makes possible the harmonic coupling of non-local systems; the aether itself is the open silently pressed string on the keyboard of the universe. By duplicating the geometry of various rotating systems that are so entangled, one will tap into the inertial energy of the aether space of those systems.* This, as will be discovered, is precisely the principle in evidence at Giza and in the Great Pyramid. The primary energies being tapped by the Pyramid are thus not nuclear, electromagnetic, or acoustic, but *the inertial energy of space-time itself.* We shall call this energy alternatively the Zero Point Energy (or ZPE), or the Scalar potential of a coordinate point.

(3) Tesla and Other Physicists React to Relativity

There was another problem with the Michelson-Morley experiment, however, and that was exposed in an experiment performed by Ernest Silvertooth in 1987.

> Einstein's special theory of relativity specifically requires that the one-way velocity of light be a constant. If that turns out not to be so, special relativity falls. The Michelson-Morley experiment, however, demonstrated only that the two-way, over-and-back average velocity of light was constant. It did not necessarily prove that the one-way velocity of light in any direction was also constant. Consequently, special relativity is founded on a tentative extrapolation that goes far beyond the experimental results of the Michelson-Morley experiment.[14]

Silvertooth's experiment demonstrated that the wavelength of a one-way beam of light varies with the direction of propagation, which strongly suggests that the one-way velocity of light can also

---

[14] Paul A. LaViolette, *Beyond the Big Bang: Ancient Myth and the Science of Continuous Creation* (Rochester, Vermont: Part Street Press, 1995), p. 250.

vary with direction, even though no one has yet satisfactorily found a way to measure such a velocity.[15]

The American Bell Laboratories physicist Herbert Ives weighed into the argument in 1938. After the publication of Sagnac's experiment, relativists, were hard pressed to explain the results, until the physicist Paul Langevin argued that the results should be adjusted to take into account the time-dilation effect of the Lorentz transforms.[16] The mental gymnastics of this assertion alone should give the reader pause, for those transforms were originally devised as "accounting tricks" to juggle the results of the *initial* Michelson-Morley experiment, the linear version, with what results should be expected if there *was* an aether! Now the transforms are being used for the exact opposite purpose, to demonstrate that there is *not* an aether and to maintain the validity of the theory *in the face of results to the contrary!*

In 1938 Ives published a paper demolishing Langevin's local time argument. And in 1951, he went further, demonstrating that "the one-way velocity of light, as defined by Einstein for a relatively moving frame, is *not* equal to a constant c as Einstein had clamed. Rather, what remains constant from one reference frame to another is a very complex mathematical function that includes readings of rods and clocks and terms describing their method of use."[17] His critique of Special Relativity and its abiding and deleterious effects on theoretical physics was devastating and thoroughgoing:

> The assignment of a definite value to an unknown velocity [the one-way velocity of light] by fiat, without recourse to measuring instruments, is not a true physical operation, it is more properly described as a ritual.... The "principle" of the constancy of the velocity of light is not merely "understandable," it is *not* supported by "objective matters of fact."
>
> With the abandonment of the "principle" of the constancy of the velocity of light, the geometries which (sic.) have been based upon it,

---

[15] Ibid.

[16] LaViolette, op. cit., p. 249.

[17] Ibid., p. 250, emphasis in the original.

with their fusion of space and time, must be denied their claim to be a true description of the physical world.[18]

In other words, the geometric assumptions provided by the Special Theory of Relativity, based as they are on the assumption of the constancy of the velocity of light, are erroneous, since they exceed the evidence adduced in support of them, or in the case of the Sagnac experiment, run counter to them.

But what of General Relativity, which seems to work so well with its geometric notion of the bending of space by large masses, and verified by the observation of the predicted bending of light around the sun? While the notion seems to be correct, its truth is perhaps less the result of observation than it is of repetitious incantation: repeat the paradigm often enough, and one will come to believe it and interpret observations on that basis. But the plain fact of the matter is that General Relativity does not explain *how* a large mass warps space; it only asserts *that* it does.

And no less than Nikola Tesla weighed in on this absurdity in 1932:

> I hold that space cannot be curved, for the simple reason that it can have no properties.... Of properties we can only speak when dealing with matter filling the space. To say that in the presence of large bodies space becomes curved is equivalent to stating that something can act upon nothing. I, for one, refuse to subscribe to such a view.[19]

The experimental death knell for General Relativity came in 1991 by scientists at Cornell. "Their computer simulations showed that if a very large oblong mass were allowed to collapse upon itself, it would produce a spindle-shaped gravitational singularity of infinite energy – a black hole – whose extremities would extend outside the black hole's central region of invisibility. Such a 'naked' singularity would radiate infinite quantities of energy into

---

[18] Ibid., p. 250, citing H. Ives, "Revisions of the Lorentz Transformations," *Proceedings of the American Philosophical Society* 95 (1951): 125-132.

[19] Ibid., p. 291, citing *The New York Times,* 10 July 1932, p. 19, col. 1.

surrounding space: an absurd result that is fatal for general relativity."[20]

E. *Non-Locality, Photon Entanglement, and Quantum States*

Nick Herbert's *Quantum Reality: Beyond the New Physics, An Excursion into Metaphysics and the Meaning of Reality* is perhaps the best one volume introduction to the subject for a layman, and I shall follow it closely here.

Let us begin with a simple question, the question, in fact, that sired the birth of quantum mechanics. Why does hot iron glow red? How would one calculate the glow of a body heated to different temperatures?

> In 1900, as the new century began, Max Planck, who against his teacher's advice had earned a degree in physics, not music, took up this black body puzzle. As a simplifying assumption he decided not to let the matter particles vibrate any way they pleased; instead he artificially constrained them to frequencies that follow this simple rule:
>
> $$E=nhf$$
>
> Where E is the particle's energy, n is any integer, f is the frequency of the particle's vibration, and h is a constant to be chosen by Planck. Planck's rule restricts the particles to energies that are certain multiples of their vibration frequency, as though energy only came in "coins" of denomination hf. Planck's constant h would later be called the "quantum of action" because it has the dimensions of energy times time, a quantity known as "action" in classical physics.
>
> .... Planck discovered that he got the same blue glow as everybody else when h went to zero. However, much to his surprise, if he set h to *one particular value*, his calculation matched the experiment exactly.... Physicists politely ignored Planck's work because although it gave the right answer, it did not play fair. Funny restriction on energy was

---

[20] LaViolette, op. cit., p. 291.

totally alien to classical physics. Newton's laws permitted particles to have any energy they pleased.[21]

The next step came with a paper by Albert Einstein on the photoelectric effect.[22]

The photoelectric effect is a simple phenomenon that is exhibited when light strikes a very thin sheet of metal: when the light strikes the metal, it knocks electrons off of the metal. Einstein demonstrated that the electrons were knocked out of the metal by the light in discrete *quanta* that corresponded with Planck's constant. And this now meant that, in addition to light's wavelike properties, it also had characteristics of particles.

> For light of a given frequency, the ejected electron's energy is always the same for the weakest light as well as for the strongest beam. When the beam is intense, more electrons come out but they all have the same energy....
>
> If you want light to give more of its energy to the ejected electron, increasing the light's intensity is not the way to do it. Instead you must increase its frequency. Light's energy evidently depends on its color, not its intensity. Blue light (high-frequency) kicks electrons harder than red light (low-frequency).[23]

These particles of light were appropriately called "photons". Thus, a new paradox was born: in some situations, light acted as a wave, and in others, like particles.[24]

The next piece of the puzzle was discovered by DeBroglie. DeBroglie essentially argued that just as light appeared to have particle properties, particles of matter "might also have wave properties."[25]   Herbert's assessment of this stage of the

---

[21] Nick Herbert, *Quantum Reality*(New York: Anchor Books, 1985), pp. 34-35.

[22] It was for this paper, in fact, and not Special or General Relativity, that Einstein won the Nobel Prize.

[23] Herbert, op. cit., p. 37.

[24] Ibid., p. 38.

[25] Ibid., p. 40.

development of the theory goes directly to the heart of the question of the underlying reality of quantum mechanics.

> It's beginning to look as if everything is made of one substance – call it "quantumstuff" – which combines particle and wave at once in a peculiar quantum style all its own. By dissolving the matter/field distinction, quantum physicists realized a dream of the ancient Greeks who speculated that beneath its varied appearances the world was ultimately composed of a single substance.[26]

But the mystery of quantum mechanics only deepens when one considers whether or not there is in fact such an underlying reality.

This is evident when one considers the relationship between the mathematical model of the theory and the reality that it models itself.

> Quantum theory is a method of representing quantumstuff mathematically: a model of the world executed in symbols. Whatever the math does on paper, the quantumstuff does in the outside world. Quantum theory must contain at least: 1. Some mathematical quantity that stands for quantumstuff; 2. A low that describes how this quantity goes through its changes; 3. A rule of correspondence that tells how to translate the theory's symbols into activities in the world.[27]

The first scientist out of the gate with such a theory was the German physicist Werner Heisenberg.

In his theory, a system of quantumstuff is represented by what is called a matrix, an example of which follows.

$$\begin{matrix} 1 & 0 & 2 \\ 3.14 & -1 & 4 \\ 0 & \sqrt{2} & -1 \end{matrix}$$

There are special rules for adding, subtracting, multiplying and dividing such matrices, the study of which is the discipline of linear algebra, an essential mathematics for so much modern

---

[26] Ibid.
[27] Ibid., p. 41.

physics. Within each matrix, there are obviously rows and columns, and each number is called an "entry". Thus, Heisenberg models quantumsuff by a *set* of such matrices, and thus his version of quantum theory is often called "matrix mechanics."

> A matrix is a square array of numbers like a mileage table on a road map which (sic.) lists the distances between various cities. Each Heisenberg matrix represents a different attribute, such as energy or momentum, with the mileage chart's cities replaced by particular values of that attribute. The matrix's diagonal entries represent the probability that the system has that particular attribute value, and the off-diagonal elements represent the strength of non-classical connections between possible values of that attribute. Thus momentum *p* of an electron is not represented by a number as in classical physics, but by one of these square arrays.[28]

Bear in mind this summing of entries on the diagonal, for it have great importance when we turn to a discussion of how the chosen mathematical model can greatly influence, and distort, the interpretation of the underlying reality.

The Austrian physicist Erwin Schrödinger, who modeled it as a wave form, proposed a second mathematical model of quantumstuff.[29] And finally, Paul Dirac symbolized "quantumstuff as an arrow (or vector) pointing in a certain direction in an abstract space of many dimensions.... A big part of Dirac's theory concerns how to change from one coordinate system to another, how to *transform* between seemingly different descriptions of the same rotating arrow."[30] It is this aspect of quantum mechanics that explains its great success and flexibility in modeling the sub-atomic world. It has a "multilingual facility" that allows physicists to choose the mathematical model most appropriate to the situation or problem they wish to explore.[31]

At this point, however, an important thing occurred in the history of science, for the theory diverged into two paths. Some

---

[28] Ibid., p. 41.
[29] Ibid., p. 42.
[30] Ibid.
[31] Ibid., p. 43.

viewed the theory merely as a means to manipulate the world. Others viewed it as a window into the deepest realities of the micro-cosmos. Which was it?

If one reflects for a moment on Heisenberg's matrix mechanics version of the theory, wherein various attributes of a sub-atomic particle such as an electron are modeled by a set of matrices, then the reality question comes into focus. Any particle of quantumstuff does not possess certain of its attributes innately. These attributes are called the dynamic attributes since they are subject to change, such as position or momentum. They seem to be "created by the measurement context itself,"[32] that is to say, by the mere act of observing them. And this raises the obvious question. If they are created by observation, or in any degree influenced by it, what then is the reality of the stuff itself? Is it real in and of itself? Or does observing it, so to speak, create reality?

The mathematician John von Neumann, about whom we will have much more to say, posed the question in his famous "proof":

> What von Neumann showed was that if you assume that electrons are ordinary objects or are constructed of ordinary objects – entities with innate dynamic attributes – then the behavior of these objects must contradict the predictions of quantum theory.... Thus, according to the quantum bible, *electrons cannot be ordinary objects, nor can they be constructed of (presently unobservable) ordinary objects.* From its mathematical form alone, von Neumann proved that quantum theory is incompatible with the real existence of entities that possess attributes of their own.[33]

But almost as soon as von Neumann had proven this, physicist David Bohm proved the opposite.

Bohm constructed a model of the electron, possessing innate dynamic attributes, which conformed to the predictions of the theory. This he did by connecting the electron to a new field, which he called a pilot wave, and which is invisible, "observable only indirectly via its effects on its electron. In Bohm's model,

---

[32] Ibid., p. 46.
[33] Ibid., p. 48, emphasis in the original.

quantumstuff is not a single substance combining both wave and particle aspects but two separate entities, a real wave plus a real particle."[34]

But there's just one problem with this model, and it is a problem highlighted by the assumption of the constancy of the velocity of light as an upper limit on acceleration. For Bohm's theory to work "whenever something changes anywhere the pilot wave has to inform the electron instantly of this change, which necessitates faster-than-light signaling. The fact that superluminal signals are forbidden by Einstein's special theory of relativity counts heavily against Bohm's model, but he was never able to rid it of this distressing feature."[35] Distressing only if, of course, one totally ignores Sagnac's rotational version of the Michelson-Morley experiment.

Bohm's model provoked yet another revolutionary paradigm shift in twentieth century theoretical physics, Bell's Non-Locality Interconnectedness Theorem. In 1964, John Stewart Bell was an Irish physicist working at the European Economic Community's accelerator in Geneva, when he decided to go on sabbatical to explore the quantum reality question.

> The first question Bell asked was: how was Bohm able to construct an ordinary reality model of the electron when von Neumann had proved that nobody could ever do such a thing? Bohm's model actually did what it claimed: it duplicated the results of quantum theory using a reality made of nothing but ordinary objects. So the fault must lie not in Bohm's model but in von Neumann's proof.
> .... As he examined von Neumann's proof, Bell wondered whether a truly ironclad argument could be constructed which would set firm limits on the sorts of realities that could underlie the quantum facts.
> ....Arguing from quantum theory plus a bit of arithmetic, bell was able to show that any model of reality whatsoever – whether ordinary or contextual – must be <u>non-local</u>. Bell's theorem reads: the quantum facts plus a bit or arithmetic require that reality be non-local. In a local reality, influences cannot travel faster than light. *Bell's theorem says*

---

[34] Ibid., pp. 48-49.
[35] Ibid., p. 50.

that in any reality of this sort, information does not get around fast enough to explain the quantum facts: reality must be non-local.

....Suppose reality consists of contextual entities which (sic.) do not possess attributes of their own but acquire them in the act of measurement, a style of reality favored by Bohr and Heisenberg. *Bell's theorem requires for such entities that the context which (sic., et passim) determines their attributes must include regions beyond light-speed range of the actual measurement site. In other words, only contextual realities that are non-local can explain the facts.*[36]

This non-locality interconnectedness theorem is yet another vital component of the physics embodied at Giza, for there two non-local systems – the solar system and the Milky Way galaxy – are quite obviously harmonically coupled in such a fashion as to suggest that inertial energy is being drawn from them. If such energy is being drawn, then it must rely on the instantaneous transfer of information (inertial energy) from the geometrical configuration of the three systems (earth, the solar system, the galactic system). The idea that reality is a non-local substrate of quantumstuff or aether has already been encountered in the previous chapter.

So what, then, *is* being measured in quantum mechanics? This question leads to the heart of the issue, the "Quantum Measurement Problem." If there is one universal force to which all objects quantum or otherwise, are subject, it is gravity. "Every object we see is continually pulsing to the gravitational rhythm of distant stars."[37] As we shall see, the Great Pyramid pulses to a great many planetary and celestial rhythms. In order to understand this problem, we need to go back to the photoelectric effect and to yet a fourth version of quantum mechanics theory, the "sum-over-histories" or "path integral" approach of American physicist Richard Feynmann. If we now modify the photoelectric experiment to shoot a beam of light through a very small aperture onto metal, electrons will be kicked off and form a pattern of

---

[36] Ibid., pp. 50-51. Underlined portions are emphasized in the original, italicized portions are added emphases.

[37] Ibid., p. 131.

concentric rings, a wave form, rather similar to the wave that results when we toss a stone into a pond.

Now quantum mechanics tells us that the dynamic attributes of an electron – position and momentum – are contextual, that is, they are created or influenced to a certain extent by the act of measurement itself. If we modify the experiment still further, and place *two* slits or apertures through which each photon of light passes, then we will see on the screen or metal (the interferometer) behind it a classical interference pattern. The problem now is *which path did the photon take?* Feynmann answered this question by saying essentially that while one may not be able to state which path an individual photon took, one may be able to "average" the paths of several photons to derive a sort of "statistical history" of the path most likely chosen.

But this idea of a statistical average does not dispense with the problem, but only sharpens it. In one version of the problem, it means that physicists cannot represent any quantum system's physical state in classical terms but rather as a "wave of possibility." *But the* **description** *of the possibility must still be couched, like any other aspect of human experience, in terms of a classical, concrete reality.* So where does the boundary between our "classical", "real" world and the quantum world lie?[38]

Now note in the two-slit experiment yet another difficulty. If we are beaming photons at the slits, the presumably, according to quantum theory, any given photon goes through one, or the other, or even *both*, slits. So why should identical quantum entities develop any differences?[39]

The Great Pyramid itself suggests one possible answer to this problem: quantum reactions are responses via Bell's non-locality theorem to quantum states in the measuring device itself. It is important to understand what is being said here. To assert that quantum states of measured systems are to some extent the result of the quantum states of the measuring system is to say, "there is a

---

[38] Ibid., p. 142.
[39] Ibid.

sense in which *atoms are made of measuring instruments* and not the other way around. As Hesienberg writes: 'Only a reversal of the order of reality as we have customarily accepted it has made possible the linking of chemical and mechanical systems of concepts without contradiction.'"[40] In other words, in pre-quantum physics, macroscopic objects such as a planet or a sun were explained in terms of the atoms that comprised them. In this new view, however, it is the other way around. Atoms and sub-atomic particles are explained in terms of the macroscopic context in which they occur.[41]

Now we take a final step. Recall that Feynmann's "path integral" or "sum-over-histories" approach to the two slit problem ultimately means that a photon takes *all* possible paths simultaneously towards its target. Re-enter John von Neumann, who posited this view as the *only* possible view of the world. The path of any particle in his view followed a "ruthless territorial imperative demanding that it occupy all its possibilities at the same time. *The fact that most of these paths are obliterated by destructive interference in no way alters a (particle's) primal orders: fill the Earth with your essence!*"[42] In other words, objects such as planets or stars or atoms in the classical sense arise as a result of the canceling out of all other alternatives. It follows then, that the proper type of interference with the waveform of those objects, i.e., with the proper harmonics, a destructive interference can be established that will simply "cancel out" or nullify the objects themselves. That is, interference can be established in an object causing all its particles to again take all paths; the object will simply appear to disintegrate in a violent cataclysm of all forms of energy.

For Bohr, this meant that an electron's attributes were relations between the electron and its measuring device. "These so-called attributes are not intrinsic properties of quantum systems but

---

[40] Ibid., p. 144.

[41] Ibid.

[42] Ibid., p. 145, emphasis added.

manifestations of 'the entire experimental situation.'" [43] That is, one may view all reality, regardless of scale, as existing in some quantum state.

F. Plasma Cosmology

We have already encountered the electromagnetic plasma cosmology of the Swedish physicist Hannes Alfvén in the previous chapter.[44] Lerner summarizes this new cosmology as follows:

> Starting in 1936 Alfvén outlined, in a series of highly original papers, the fundamentals of what he would later term cosmic electrodynamics – the science of the plasma universe. Convinced that electrical forces are involved in the generation of cosmic rays, Alfvén pursued ...(a) method of extending laboratory models to the heavens.... He knew how high-energy particles are created in the laboratory – the cyclotron, invented six years earlier, uses electrical fields to accelerate particles and magnetic fields to guide their paths. How, Alfvén asked, would a cosmic, natural cyclotron be possible?
> ....But what about the conductor? Space was supposed to be a vacuum, thus incapable of carrying electrical currents. Here, Alfvén again boldly extrapolated from the lab. On earth even extremely rarified gases can carry a current if they have been ionized – that is, if the electrons have been stripped from the atoms....Alfvén reasoned that such plasma should exist in space as well.[45]

This may not sound too revolutionary, until one notices what is unique about this theory: "certain key variable do *not* change with scale – electrical resistance, velocity, and energy all remained the same. Other quantities *do* change: for example, time is scaled as

---

[43] Ibid., p. 161. The idea of relational attributes is not new to Bohr, incidentally, but is to be found also in Thomas Aquinas, once again highlighting the deep and profound connection between religion, metaphysics, and physics. Cf. Aquinas, *Summa Contra Gentiles*, Pt. 4, Question 2, Art. 5.

[44] For a lucid and provocative account of this cosmology and the issue of the relationship between theory, observation, and experiment that it raises, cf. Eirc J. Lerner, *The big Bang Never Happened* (New York:Vintage Books, 1992).

[45] Lerner, op. cit., p. 181.

135

size, so if a process is a million times smaller, it occurs a million times faster."[46]   In other words, the whole difficulty of post-relativistic physics, that of reconciling relativity with quantum mechanics is avoided entirely. And notice what the principle differential is.   The principle differential is exactly what the examination of paleophysics in the previous chapter said it would be, time. But the other laws operate no matter what the scale.[47]

Since time is scale-sensitive, and other electromagnetic forces are not, the implication is revolutionary.

> Equally important, though, is the converse use of these scaling rules. When the magnetic fields and currents of these objects are scaled down, they become incredibly intense – millions of gauss, millions of amperes, well beyond levels achievable in the laboratory. However, by studying cosmic phenomena, Alfvén shows, scientists can learn about how fusion devices more powerful than those now in existence will operate. In fact, they might learn how to design such devices from the lessons in the heavens.[48]

Notice the coupling of electromagnetic vorticular processes with fusion itself, a process duplicated in the work of Philo Farnsworth's plasmator patent (cf. below), which uses virtual electric fields to establish a stable fusion reaction in a cloud of plasma or ions.

But notice something else. Lerner clearly implies that *if the inertial and electromagnetic processes of the heavens are somehow captured, i.e., if one can couple to them, "fusion devices*

---

[46] Ibid., p. 192, emphasis added.

[47] Lerner, op. cit., p. 193: "In the reedition of *Cosmic Electrodynamics,* written with Fälthammer in 1963, Alfvén gives filamentation a central role in producing homogeneities in plasma, on scaled from laboratory up to stellar nebulae – the vast clouds of glowing gass surrounding many star clusters in a galaxy. When a current flows through a plasma, Alfvén shows, it must assume the form of a filament in order to move along magnetic field lines." This is to say that it is possible to produced spiral filamentary patterns in highly rarified plasma in the laboratory. This has not only been done, but the spiral patterns exhibited almost exactly duplicate the patterns of various galaxies that have been observed. "As above, so below."

[48] Ibid., pp. 192-193.

more powerful than those now in existence will operate." What fusion devices could he be talking about? No tokamak magnetic bottle has ever achieved a stable controlled fusion reaction, and it is unlikely that Lerner knows about Farnsworth's plasmator, since he nowhere mentions it (even though it would appear to be based upon the same theoretical assumptions). The only thing left, then, are the city-busting superbombs that fill French, American, and Russian arsenals.

The coupling to the heavens is, of course, what one finds in the structures at Giza, and in the Great Pyramid itself. And as will be seen in the final chapter, there is very good reason to assume that it utilized precisely the electromagnetic properties of plasma in exactly this fashion.

Alfvén suggested other aspects of plasma cosmology in an article published in 1942.

> If a conducting liquid is placed in a constant magnetic field, every motion of the liquid gives rise to an (electromagnetic field) which produces electric currents. Owing to the magnetic field, these currents give mechanical forces which (Sic.) change the state of motion of the liquid. *Thus a kind of combined electromagnetic-hydrodynamic wave is produced which, so far as I know, has as yet attracted no attention.*[49]

This, as we shall see in the next chapter, sounds very much like the electro-acoustic waves discovered by Tesla in his high frequency direct current impulse experiments.

But there are even more profound resemblances between the aether of paleophysics and the modern plasma cosmology, not the least of which is the concept that the universe exhibits a filamentary and *cellular* structure. Not only does the universe exhibit "electric layers" of various densities like a fluid, but

> Cosmic plasmas are often not homogeneous, but exhibit *filamentary structures* which are likely to be associated with currents parallel to the magnetic field.... In the magnetospheres there are thin, rather stable

---

[49] Hannes Alfvén, "Existence of Electromagnetic-Hydrodynamic Waves," *Nature*, No. 3805, October 3, 1942, pp. 405-406.

current layers which separate regions of different magnetization, density, temperature, etc. It is necessary that similar phenomena exist also in more distant regions. This is bound to give space a *cellular structure* (or more correctly, a cell wall structure). [50]

In this article, Alfvén makes his case for an unequal (inhomogeneous) distribution of matter in the universe by pointing out that there is also an upward limit for the size of things in the universe, a limit called the Laplace-Schwarzschild limit, representing an "instability limit."[51] His comment on this limitation is worth citing: ""his instability cannot be due to a release of nuclear energy - as in stars - because for the large structures we consider this to be insufficient. Hence, if we do not want to introduce new laws of nature, there are only two energy sources available: *gravitation* and *annihilation.*"[52] As we shall also see, there is the possibility that gravitational force was somehow also being accessed and manipulated in the Great Pyramid.

The regions of the universe devoid of matter suggested to scientists that it was "clumpy." And this may be one of the most profound insights of plasma cosmology. Alfvén's comment must be cited, for it deserves extensive commentary.

This means that stars should be organized in galaxies G_1, a large number of these galaxies form a larger 'galaxy of type G_2. – we would today prefer to speak of a 'cluster' – a large number of these a still larger structure G_3, and so on to infinity. Chartier showed that the mean density of a structure of size R must obey the relation

$$p \sim R^x$$

with x › 2. This leads to an infinite universe with infinite mass but *with an average density zero.*[53]

---

[50] Hannes Alfvén, "On Hierarchical Cosmology," *Atsrophysics and Space Science,* Vol. 89, (Boston: D Reidel, 1983), 313-324, p. 314.

[51] Ibid., p. 318.

[52] Ibid., p. 319, emphasis in the original.

[53] Ibid., p. 316.

What does this mean?

Recall that our discussion of matrix algebra indicated that there were special rules for adding, subtracting, multiplying and dividing such matrices. Imagine now that one is representing the average mean density of matter in the universe by such a matrix. It would, of course, be a very complex procedure, but nonetheless, if one summed all the entries in the matrix, one would end up with a sum of zero. This is called a "zero-summed" matrix, and it would represent what an observer external to the universe – God - would see in terms of its mean density. Now, we know from simple arithmetic that any number times zero equals zero. And the same holds true in linear or matrix algebra, where such a number is called a "scalar", representing in physics a pure "magnitude of force" but without any *direction*.

This is a very important consideration, because one can *also* represent any point in vacuum space by a similar such matrix. Imagine you are holding a rubber ball in your hand, squeezing it. There is force *inside* the ball, but none observable *externally* to the observer. Now imagine you have mathematically modeled that squeezed rubber ball by a zero-summed matrix. The force you are loading into the ball – the squeezing – is the scalar. But since modern physics uses this matrix mathematics to model so much, it will tell you that there is no force present at the point.

However, when Maxwell first modeled the equations of electromagnetism, he did *not* use this type of mathematics. Rather, he used a form of mathematics called "quaternions" which said that a scalar – the squeezing of the ball – times a zero-summed matrix returned not zero, but the scalar. This meant he employed a form of mathematics that looked at the energy locked up *inside* a system, rather than *outside*. Notice also that Alfvén, in our rather crude analogy, is saying almost the same thing. The implication is clear; for plasma cosmology, there is a uniform way of looking both at micro-sized and macro-sized objects in the universe. "The same general laws of plasma physics hold from laboratory and magnetospheric and heliospheric plasmas out to interstellar and intergalactic plasmas."[54] We shall have more to say about this "scalar" physics in a later section.

The cellular structure of the cosmos suggested by plasma cosmology strongly parallels the paleophysical view of the universe as a "living" organism: when something happens in one place, the whole reacts. But how is this possible? Bell's theorem demonstrated that reality is non-local. Thus, if one assumes the existence of an aether, as a "field of information", it is rather easy to see how what happens in one "cell" of the universe is quickly transmitted to another, for all the cells are "entangled" or "coupled" to each other.

G. *Scalar Interferometry and Nonlinear Optical Phase Conjugation*

Scalar interferometry and phase conjugate waves, non-equilibrium thermodynamics, and sub-quantum kinetics are the next components of physics that are needed to understand the Great Pyramid's possible functioning as a weapon. These areas have been developed only in the last few years, and one can only speculate on what research is being done largely in secret. But what has been published is tantalizing enough. Let us begin with phase conjugate waves and scalar interferometry.

In our squeezed rubber ball analogy, we discovered how it was possible to lock energy *inside* of something that would still look to an external observer as if there was no energy inside it. Now imagine alternatively squeezing and releasing the rubber ball, or "pulsing" it. This would produce a wave moving back and forth inside the ball.[55] If one now imagines that the ball is the universe itself, then it would seem to follow that all parts of the ball, the "cells", react to the pulsing immediately. This would imply that the "wave" inside the ball is somehow faster than light.

---

[54] Hannes Alfvén, "Cosmology in the Plasma Universe," *Laser and Particle Beams* (Cambridge: Cambridge University Press, 1983) 389-398, p. 389.

[55] Cf. V.K. Ignatovich, "The Remarkable Capabilities of Recursive Realtions," *American Journal of Physics,* Vol. 57, No. 10, 873-878, p. 874.

But is such faster-than-light transference of information possible? Indeed it is. The French physicist Alain Aspect performed yet another version of the light-beam splitting experiment designed to test Bell's Non-Locality theorem in which photons of light were given certain wavelength characteristics called "polarization", and then the split beams of light were then measured at various places and distances. Remarkably, the photons continued to show evidence of their original coupling, and distance was not a factor.[56]

This implies that the split photons somehow reacted to each other by some "non-linear" means. This means would appear to have something to do with "nonlinear optical phase conjugation."[57] This rather difficult sounding phrase is really rather simple to understand. If one shines a flashlight at an ordinary mirror at an angle of 45 degrees from its surface, we know that the beam will be reflected from the mirror at the same angle away from the mirror, giving a ninety degree angle between the incoming and outgoing beams. But the reflected beam will scatter and widen, not only according to the normal inverse square law of light propagation, but also because the mirror is an irregular surface. It is not "in phase" with the wavelength of the light, so to speak.

But imagine that one could construct a mirror that was so harmonically attuned to the incoming beam of light that it reflected the light back in the same direction it came, perfectly, and without any scattering. The reflected beam not only would *not* obey the inverse square law, but it would in effect be a "time-reversed" light wave, the exact opposite of the incoming beam. David M. Pepper comments on one possible use of such "phase conjugate mirrors" that are able to adjust the wavelength and time of split light beams:

---

[56] Alain Aspect, Philippe Grangier, and Gérard Roger, "Experimental Tests of Realistic Local Theories via Bell's Theorem," *Physical Review Letters*, Volume 47, Number 7 (17 August 1981), 460-463. "No effect of the distance between measurements on the correlations was observed." (p. 463)

[57] Cf. David M. Pepper, "Nonlinear Optical Phase Conjugation," *Optical Engineering: the Journal of the Society of Photo-Optical Instrumentation Engineers,* Vol. 21, No. 2, (Mar-Apr 1982), 156-183.

(A) parallel amplifying system could be used to initiate fusion. A fusion pellet is illuminated by a pulsed low-power laser... Pulses scattered from the pellet pass through the three parallel laser amplifiers ... The intensity of the individual pulses is increased, but at the expense of introducing distortions. The pulses are then directed to a four-wave-mixing phase-conjugate mirror. By the time they reach the mirror they are out of step because they have traveled different distances. When all the pulses are inside the mirror, it is turned on, conjugating each pulse and reversing their temporal order with respect to one another... On the return trip the distortions are removed and the pulses become synchronized, so that an intense pulse of radiation hits the pellet.[58]

There are some features here that are quite crucial to understanding how the Great Pyramid might have functioned as a weapon.

First, there is a coherent energy output (the laser). A beam of laser light is called "coherent" because all its photons march in lock step or "in phase", like a troop of solders. This gives laser light its extraordinary power. When soldiers march across a bridge, they intentionally break step, since if they continued to march in phase with each other, they would set up a vibration in the bridge that could – and often did – collapse the bridge because energy was loaded into the bridge before it could dissipate.

Second, there is a target, in this case, a "pellet" of material capable of undergoing a thermonuclear fusion reaction, in this case perhaps it would be a capsule of lithium deuteride. As we know by now, however, *every* object has a frequency that it vibrates or resonates to. Thus, if one knew the frequency of any material, with proper "phase conjugation" of energy, one could conceivably initiate a nuclear reaction by loading enough vibrations into it, by causing it, in other words, to vibrate or "cavitate" violently enough to explode.

---

[58] David M. Pepper, "Applications of Optical Phase Conjugation," *Scientific American*, Vol. 254, No. 1 (Jan 1986), 74-83, p. 82.

Third, there is a *split* beam whose parts all arrive at the target "in phase" from several angles, increasing the energy loaded into the area (remember our squeezed rubber ball?).

Finally, there is the phase-conjugate mirror, which collects the split beams and fires them on to the target in such a way that all beams are not only time-reversed but so that they arrive at target at the same time and exactly in phase with each other. The time-reversed component is essential because, it will be recalled, it means that the beams do *not* suffer the effect of dissipation due to the inverse square law. These waves would arrive at the target and set up a "standing wave" inside it, rather like a sound wave, causing it to vibrate in lock step with the incoming pulses of energy. Eventually, the energy locked inside the target would cross a threshold of stability, and it would blow itself apart *in a nuclear reaction regardless of the material that comprised the target.*

That such electromagnetic induction of thermonuclear reactions – the same reactions that power the sun and hydrogen "super-bombs" – is possible will be explored in the section on cold fusion. Of course, such engineering capabilities are beyond our current technology, but it takes little imagination to envision a very large-scale application of Pepper's idea. As we shall see in chapter five, there is very strong evidence that the Great Pyramid employed a large maser as one component of its energy output.[59] And there is one more consideration. Since every object has its own resonance, and yet is part of the same universe, this means to some extent that it is vibrating sympathetically with every other object in it, particularly with those objects in its immediate location – star systems, galactic system, and so on. To initiate the type of nuclear reaction in such an object, one would perforce have

---

[59] A "maser" is exactly like a laser, only using the invisible microwave part of the electromagnetic spectrum rather than visible light. The power of such a device can best be imagined by remembering that an ordinary kitchen microwave oven is like a microwave version of a light bulb; its photons do not march in lock step like a company of soldiers, but scatter like the ordinary light from a lamp.

to take into account the harmonics of those systems in which the target is found. This, of course, is exactly what one discovers in the Great Pyramid. In this respect, it constitutes a gigantic and *super-sophisticated* "phase conjugate mirror," picking up the incoming inertial vibrations of the aether as they have been "split" into the acoustic, electromagnetic, nuclear and gravitational vibrations of the earth, the solar system, and the Milky Way Galaxy, and modulating its output energy by the same vibrations in a deadly carrier wave of unheard of destructive potential. In chapters five and six, we will present evidence that suggests that this type of phase conjugation may very well have been the function of three of its better-known internal chambers.

H. And Some Very Strange Ideas, Patents, and Experiments

Physics, besides suffering these internecine conflicts within its "orthodox" confines, also possesses persistent stories and allegations of even more bizarre experiments that seem to run counter to received theory. Here we shall quickly summarize some of them that I believe to be relevant to the weaponization of the hyperdimensional paleophysics in the Great Pyramid.

(1) Endothermic, or "Cold" Fusion

In March of 1989, two physicists at the University of Utah held a press conference to announce a discovery that shook the edifice of theoretical and experimental physics to its very foundations. "It proclaimed that Professors Martin Fleischmann and Stanley Pons had discovered that practical nuclear fusion could be carried out with relatively simple laboratory equipment at near room temperature – called the Fleishmann-Pons Effect (FPE). The great news was that the experiment could *produce more energy output than was required as imput.*"[60] Pons and Fleischmann achieved

---

[60] Hal Fox, *Cold Fusion Impact in the Enhanced Energy Age* (Salt Lake City, Utah: Fusion Information Center, 1992), p. I-1, emphasis in the original.

their result by modifying the simple and well-known process of electrolysis. Gone were the multi-million dollar Tokamak wiggler magnets, gone the unstable hot plasmas.

Yet, despite numerous successful reduplications of their experiment, there was a problem. *Some* experiments failed to reduplicate their results. There seemed to be no predictability or regularity – at least none known or presently quantifiable – to the result. As a consequence, both scientists were literally hounded out of the country by a more "orthodox" scientific establishment. But we perhaps already have a clue: *inadequate attention was paid to the overall geometrical and therefore harmonic configuration of the experiments.* This has an important, and often over-looked, corollary. During the first hydrogen bomb tests, the actual energy yield of the bombs far exceeded those initially calculated. There was an "x" factor, *an unknown source of surplus energy* that was being tapped. Since hydrogen bombs unleash enormous amounts of destructive energy at the very sub-atomic level of the nucleus of atoms, we may also surmise, in part, where that energy came from and why, for such weapons literally cause a violent local disturbance in the geometry and fabric of space-time. In short, *some as yet inadequately understood laws of harmonics produced the excess energy.*[61]

But the results of Pons' and Fleischmann's experiment are sufficiently credible to state the relevance of their discovery for our purposes. *It is possible to induce stable endothermic or cold fusion reactions at much lower levels of energy input than previously thought possible and it is possible to do so by ordinary electromagnetic means.* It is a short step to the conclusion that it is also possible to induce *unstable* thermonuclear reactions by a similar method.

---

[61] We may speculate further that this idea is one of the most closely held secrets of the nuclear powers, and may explain why recent additions to the "nuclear club" such as France, China, and India insist on continued testing of their weapons; they are trying to discover what those laws are.

(2) Farnsworth's Plasmator

Long before Pons and Fleischmann performed their experiments, however, another scientist may have beaten them to the punch by pursuing an entirely different methodology. His name was Dr. Philo Farnsworth. Most people are unfamiliar with his name, and yet they spend many hours every day sitting before his most famous invention being entertained: television. Farnsworth almost single-handedly conceived of the idea of television, and then proceeded to invent every component that made it possible. A glance at his many elegant patents in the U.S. Patent Office will convince one, if nothing else, that Farnsworth was an experimental and applied scientist and engineer of the highest caliber. By the end of his life it is safe to say that no one had a more thorough knowledge of what could and could not be accomplished with vacuum tubes. After all, he had designed and patented almost every tube that made television possible.

It was that knowledge that led him, toward the end of his life, into a different area of research: the control of thermonuclear fusion reactions. In two breathtaking patents he outlined a method for controlled hot fusion: patent number 3,258,402, entitled "Electric Discharge Device for Producing Interactions Between Nuclei," and patent number 3,386,883, "Method and Apparatus for Producing Nuclear-Fusion Reactions." The last patent is the culmination of his life's work.

What, exactly, did he claim and accomplish in these two patents? First, like Pons and Fleischmann, he avoided entirely the "need for the gigantic"[62] that characterizes so much "official"

---

[62] Gerry Vassilatos, "The Farnsworth Factor: The Most Notably Forgotten Episode in 'Hot' Fusion History," *Borderlands*, Second Quarter, 1995, p. 2. It is worth mentioning that Vassilatos has his own version of "paleophysics" research, a research that concentrates on the forgotten experiments and observations of the last two hundred years: "Old texts preserve forgotten thoughts...not disproven thoughts....Discovery and anomaly are rare gifts which (sic., et passim) must be honored and preserved until understood. The scientific

corporate and government sponsored fusion research. Using electron optical focussing to concentrate ions in what he called a "fusor" tube, comprising a spherical anode surrounding a cathode. On the anode portion of the electron tube, ion cannons were mounted so that their beams would intersect in the center of the cathode. This established "in the cathode interior a series of concentric spherical sheaths of alternating maxima and minima potentials," or, in effect, *virtual* electrodes. As a result, the ions that were captured "in the center-most virtual electrode have fusion energies and are contained at a density sufficient to produce fusion reactions."[63] By 1965 Farnsworth had tested and achieved stable reactions for over thirty seconds, in a design of an electron tube no larger than a softball and that solved simultaneously both the conversion and containment problems that have dogged the multi-million dollar failures of official research.

What happened to Farnsworth's device and patents? ITT, which had helped to fund the research, bought the patents outright, and suppressed any further development along the lines of containment via virtual electrodes. The line of research pioneered by Farnsworth, *as far as we know*, was allowed to fade away from view.

(3) The Sonoluminesence Effect

Between the world wars a group of German physicists discovered a one of those curious phenomena that seem

---

historian methodically searches out catalogues of forgotten phenomena by thorough examination of old periodicals, texts, and patent files. The retrieval of old and forgotten observations, discoveries, scientific anecdotal records, and rare natural phenomena provide the intellectual dimension desperately needed by modern researchers who work in a vacuum of dogma.... The trained researcher identifies, distinguishes, and secures those particular forgotten discoveries which violate contemporarily held theoretical models. The aim of this research is new knowledge through reevaluation." (p. 4) Vassilatos is unparalleled in his ability to analyze the motivations for the suppression of certain types of technology or scientific theory. (Cf. p. 12).

[63] Ibid., pp. 7-8.

perpetually to be shuffled to the sidelines of scientific research. Only in this case, we have clear evidence that the "shuffle" took the research straight into the labyrinth of the bizarre and very secret wartime research of the Third Reich. These scientists discovered that sound waves traveling in water could produce bubbles, and could then be used to vibrate or cavitate the bubbles to the point that they suddenly burst into miniature explosions of blue light. They called the phenomenon "sonoluminesence".

The phenomenon was not fully explained until the 1990s when American physicists theorized that the burst of blue light was caused by a sonic boom taking place *inside* the region of the bubbles. Once again, we are back at the analogy of the squeezed rubber ball. What produced the light was compressed air as it was heated to more than several hundred thousand degrees, a temperature hotter than the sun's surface. The heat produced a plasma, in other words, the intense temperatures of which tore electrons loose from their atoms and producing billions of charged particles and giving off electromagnetic radiation in the form of the blue light.

These scientists made a startling conclusion – startling to modern science, but not, perhaps, to the paleophysics already encountered – that there was a relationship between acoustic and electromagnetic energy that was more intimate than previously thought to be the case. This leads to a bold conjecture on the possible weaponization of such a phenomenon.

(1) Every object or material substance has a resonant frequency.

(2) Sound waves that are resonant to that frequency can be amplified.

(3) Once amplified, they can be projected and accelerated to that object by electromagnetic forces – rather like a radio wave carries, or is "modulated" by acoustic information.

(4) Once the carrier electromagnetic wave reaches the target, the acoustic energy enters inside of the target, resonating harmonically with it, and loading energy into it.

148

(5) The threshold of instability is reached and the target is disrupted at the sub-atomic level by breaking down the bond between the sub-atomic particles.

(6) It is most effective if the carrier beam as well as its acoustic modulation is itself resonant with the target.

If one now recalls all that has been said thus far about phase conjugate mirrors, the coupling of non-local systems and the various forms of energy – gravitational, acoustic, electromagnetic – as well as time-reversed phase conjugation, *and engineers these components together,* one comes very close to an understanding of the awesome weapon the Great Pyramid was, and how it worked, for as we shall discover in chapter five, not only did it have an electromagnetic cohered output in the form of a maser, but also very likely an acoustic amplification and modulation of that output. One may speak of it as being a phase-conjugate electro-acoustic "howitzer."

(4) Eastland's Arco Patents

There exists at Gakona, Alaska a military and research installation known as the High Altitude Auroral Research Project, or HAARP, for short. The purpose of this vast phased antenna array is to beam billions of watts of power up into the ionopshere of the earth, thus "heating" a portion of the atmosphere and "lifting" it, creating an area of low pressure. To understand the relevance of this frightening project to the Great Pyramid weapons system, we need only understand two things about it. First, it was conceived originally as a component of the Strategic Defense Initiative project during the Reagan-Bush administrations.

Second, and much more important, are the original patents by American physicist Bernard Eastland on which the project is based, U.S. Patent number 5,038,662, "Method for Producing a Shell of relativistic Particles at an Altitude Above the Earth's Surface," patent number 4,712,155, "Method and Apparatus for Creating an Artificial Electron Cyclotron Region of Plasma." But patent number 4,686,605 interests us most. Its principal design is to heat the ionosphere, are listed the following in its abstract:

> A method and apparatus for altering at least one selected region which normally exists above the earth's surface. The region is excited by electron cyclotron resonance heating to thereby increase its charged particle density. In one embodiment, circularly-polarized electromagnetic radiation is transmitted upward in a direction substantially parallel to and along a field line which extends through the region of plasma to be altered....This increase in energy can cause ionization of neutral particles which are then absorbed as part of the region, thereby increasing the charged particle density of the region.[64]

But what practical application would such heating accomplish? The patent speaks for itself: disruption of land, airborne, sea and surface and sub-surface communications, missile or aircraft destruction, deflection, or confusion, weather modification, and, since HAARP broadcasts within the frequency of the human brain, behavior modification.[65]

By heating a portion of the earth's ionosphere and building up a massive charge above a certain region, a threshold of instability would be obtained, an electrical potential differential between the region of particles above the earth, and the earth's surface itself. Selecting the region is simple: one merely configures the antennae to be sufficiently out of phase to guide the interference signal to the point above the planet's surface one wishes to "heat". Converging there, the signals supercharge the ionosphere, causing the charge differential, which would eventually be discharged to the surface in several massive bolts of electricity per second, each bolt *far exceeding anything in the largest natural thunderstorm.* A surface target would be obliterated in an electrical blast much like a large H-bomb without the collateral radioactivity of such bombs.

I cite these patents, however, not to make any wild speculations about HAARP nor its potentialities. My purpose in citing them is to exhibit the type of *thinking* currently being exhibited by the

---

[64] Abstract, U.S. Patent Number 4,686,605.
[65] Ibid.

American military. A weaponized technology is being created that is substantially *unified*. That is, one installation can be used for a variety of communications, defensive, or offensive military purposes. The hardware of the installation does not change. Rather, its harmonics are adjusted to accomplish whatever purpose is desired at the moment.

(5) The Philadelphia Experiment

The Philadelphia Experiment has about it the stuff of legends. Four prominent American physicists were said to be involved: Thomas Townsend Brown, John von Neumann, Albert Einstein, and Nikola Tesla. Allegedly, the experiment took place in 1943 on the *DE* (destroyer escort) *USS Eldridge* in the Philadelphia navy yard. To understand it, we need to recall that at the beginning of World War Two, the German navy used underwater magnetic mines and torpedoes that were attracted to the magnetic field generated by large metal ships. The mines would literally be attracted to the hulls of Allied ships, explode, and sink them.

The Allies quickly learned how to counter these devices through a technique called "degaussing." A large electrical coil began to be fitted around the outside hull of ships, through which a current was passed that was exactly out of phase with the magnetic resonance of the ship. But here is the clincher. Since the ships themselves moved *through* and in the earth's magnetic field, and this varied minutely from place to place, the harmonic resonance of any individual ship also varied as it moved from one place to another. Thus, the resonance in the degaussing coils on the ships had to be monitored and constantly adjusted at all times to make sure that the ship was not "magnetically visible" to the German mines and torpedoes. The slightest imperfection in this resonance would leave a large magnetic signature, and the ship would be "visible."

The experience with degaussing suggested a bolder defense. Einstein had already published his theory of General Relativity, which maintained light could be bent in a strong gravitational field, and he was working on a "Unified Field Theory" which would explain both electromagnetic and gravitational energy. If ships could be made magnetically invisible, could they not also be made invisible to radar?

Thus was born the Philadelphia Experiment. Using not one, but *three* coils around each axis of the ship, and adjusted to be in phase with one another, the idea was to create a rotating electromagnetic field around the ship that would literally bend a radar signal around the ship altogether. The signal would flash by the ship without a return to the German radar sets, and the German operators would conclude no ship was there. The experiment makes rational sense in wartime America, with the invasion of Europe impending.

What is actually alleged to have happened, however, was more than the scientists bargained for, regardless of which version of the story one reads. When the coils were turned on and the field was established, immediately there was a loud electrostatic "buzz" in the air, and a green mist soon surrounded the ship. This point actually serves to substantiate what happened next, for this green mist many will recognize as simply the ionization of normal air that takes place in the presence of strong electromagnetic fields. Anyone familiar with tornadic thunderstorms in the American midwest will recognize this peculiar green color of ionized atmosphere that precedes a violent storm.

Once the green mist lifted, however, the ship was not just invisible to American radar operators, it was invisible to the naked eye. In one version of the story, all that one could see was the outline of the hull in the water. In another version, the ship was simply totally *gone*, having reappeared some hundreds of miles away in a cloud of green mist before the confused British captain of an aircraft carrier, who made a note of it in his log.

The story takes an even stranger turn when the field was turned off. Nearly every account of the experiment states that the crew of the ship who were outside on the decks felt an overwhelming electrostatic *pressure* pressing against them, a phenomenon similar to what Tesla recorded in his high frequency direct current impulse experiments. But more importantly, several crewmembers were embedded in bulkheads and various surfaces of the ship, their bodies parts inexplicably *fused* with the metal of the ship in some grizzly sort of organo-metallic nightmare. Some crewmembers simply had to be put out of their misery, and others lost limbs.

This led the project's alleged supervisor, John von Neumann, back to the drawing boards to figure out what went wrong. Before turning to that, we note another important set of clues: ionized plasma, electrostatic pressure fields, and a "bubble" of invisibility.

(6) Montauk and von Neumann's Time Locks

John von Neumann, so the story goes, went back to the drawing board to figure out why the Philadelphia Experiment was simultaneously to stupendously successful and such a colossal failure. Von Neumann allegedly concluded that by utilizing such massive rotating electrostatic fields that the ship and its crew had somehow been "teleported" into an alternative space-time and come back again, with the result that some of the men did not come back as "in phase" as they had left, hence their embedding in bulkheads and decks. The version of the Philadelphia Experiment that interests us here is that given by "Commander X":

> Four (of the crew) who had moved from their original position wound up in the steel deck. While the ship's fields were up there was no problem. When the fields collapsed and their time locks were gone, if they had, unfortunately, changed position, drifted and rematerialized in our dimension...at a slightly different place or were unfortunate enough to be where the steel deck was, the steel of the deck literally melded with the molecules of their body....
>
> Now, the question was, what do you do to prevent it? The problem entailed understanding, which von Neumann had to go back and do, basic metaphysics. Can you imagine a hardheaded Dutchman[66] steeped in mathematics and with a particularly tough materialistic mind

---

[66] Von Neumann was actually Hungarian.

153

suddenly having to approach metaphysics? ... Nevertheless he did his homework. And he found the problem. [67]

What was this solution?
"Commander X" calls them "time locks."

> Every human being that is born on this planet, actually from the time of conception on, has what is called a set of time locks. The soul is locked to that point in the stream of time...so everything flows forward at a normal rate of flow at the time function.... These locks stay with you your entire life.[68]

While "Commander X" is not very clear, he seems to be saying that the geometrical configuration of the system into which one is born – a by-gone age would have called it your Zodiacal "sun sign" – somehow creates a "temporal harmonic" within each person with that configuration, an interpretation that would seem to be born out by his suggestion that the crew of the Eldridge that had moved out of their original position were in danger once the field was turned off. A field that literally bends light sufficiently to make an entire ship invisible can be looked at another way, since it might equally have been intense enough to send the ship into a space not in phase with our own. If one moves as it is coming back into our own space, one might, indeed, find oneself embedded in a bulkhead.

But for all that, if this was von Neumann's actual solution to the problem, then it has all the hallmarks of astrology. However, so did an RCA study of sunspots in the 1950s.

---

[67] Commander X, *The Philadelphia Experiment Chronicles: Exploring the Strange Case of Alfred Bielek and Dr. M.K. Jessup* (Wilmington, Delaware: Abelard Publications, 1994), p. 68.

[68] Ibid., p. 67.

(7) The RCA Study of Sunspots

Ham radio operators and those in the telecommunications industries know that there are periods when sunspot activity interferes more with communications than at other times. This fact prompted the RCA Company to commission a study of why this is so. The man commissioned to do the study was John A. Nelson. And what he discovered astonished both the RCA company and the scientific world. In an article entitled "Planetary Position Effects on Short-Wave Signal Quality," Nelson released the results of his study.

> To his surprise Nelson soon specifically correlated this rising and falling radio interference with not only (Sic.) sunspot cycle, but with the solar system: he found, to his increasing astonishment, a very repeatable – in essence, *astrological* correlation – between the inexorable orbits of all the planets (but especially Jupiter, Saturn, Uranus and Neptune – which, remember, hold essentially all the solar's (Sic.) system's known *angular momentum*)...and major radio-disturbing eruptions on the Sun! ...In essence what John Nelson had *rediscovered* was nothing short of...the ultimate, very ancient, now highly demonstrable *andular momentum* foundations behind the *real* influences of the Son and planets on our lives.[69]

I.   Some Conclusions:

Gathering all these thoughts together, one can glean some of the speculative principles on which the Great Pyramid might have functioned as a weapon.

(1) With the proper type of interference with the waveform of those objects, i.e., with the proper harmonics, a destructive interference can be established that will simply "cancel out" or nullify the objects themselves. That is, on one view of quantum mechanics, interference can be established in

---

[69] Richard C. Hoagland, "Hubble's New Runaway Planet-Part III," (www.enterprisemission.com), July 18, 1999, p. 2.

an object causing all its particles to again take all paths; the object will simply appear to disintegrate in a violent cataclysm of all forms of energy.

(2) If the inertial and electromagnetic processes of the heavens are somehow captured, i.e., if one can couple to them, then as Lerner stated, "fusion devices more powerful than those now in existence will operate."

(3) The cellular structure of the cosmos suggested by plasma cosmology and by the "living" universe of paleophysics suggests when something happens in one locality, the whole reacts but means of some mechanism not yet adequately comprehended. But how is this possible? Bell's theorem demonstrated that reality is non-local. If one assumes the existence of an aether, as a "field of information", it is rather easy to see how what happens in one "cell" of the universe is quickly transmitted to another, for all the cells are "entangled" or "coupled" to each other. There as thus as yet inadequately comprehended laws of harmonics that would appear to be scale invariant, and these laws would seem to indicate a unification of the various fields of physics.

(4) As we know by now, however, *every* object has a frequency that it vibrates or resonates to. Thus, if one knew the frequency of any material, with proper "phase conjugation" of energy, one could conceivably initiate a nuclear reaction by loading enough vibrations into it, by causing it, in other words, to vibrate or "cavitate" violently enough to explode.

(5) Finally, we encountered the idea of the phase-conjugate mirror, which collects the split beams and fires them on to the target in such a way that all beams are not only time-reversed but so that they arrive at target at the same time and exactly in phase with each other. The time-reversed component is essential because, it will be recalled, it means that the beams do *not* suffer the effect of dissipation due to the inverse square law. These waves would arrive at the target and set up a "standing wave" inside it, rather like a

156

sound wave, causing it to vibrate in lock step with the incoming pulses of energy. Eventually, the energy locked inside the target would cross a threshold of stability, and it would blow itself apart *in a nuclear reaction regardless of the material that comprised the target.*

In this respect, the Great Pyramid might have constituted a gigantic and *super-sophisticated* "phase conjugate mirror," picking up the incoming inertial vibrations of the aether as they have been "split" into the acoustic, electromagnetic, nuclear and gravitational vibrations of the earth, the solar system, and the Milky Way Galaxy, and modulating its output energy by the same vibrations in a deadly carrier wave of unheard of destructive potential. In chapters five and six, we will present evidence that suggests that this type of phase conjugation may very well have been the function of three of its better-known internal chambers.

As we also discovered, this idea has an important, and often over-looked, corollary. During the first hydrogen bomb tests, the actual energy yield of the bombs far exceeded those initially calculated. There was an "x" factor, *an unknown source of surplus energy* that was being tapped. Since hydrogen bombs unleash enormous amounts of destructive energy at the very sub-atomic level of the nucleus of atoms, we may also surmise, in part, where that energy came from and why, for such weapons literally cause a violent local disturbance in the geometry and fabric of space-time. In short, *some as yet inadequately understood laws of harmonics produced the excess energy. Thus, it might be possible to induce stable endothermic or cold fusion reactions at much lower levels of energy input than previously thought possible and it is possible to do so by ordinary electromagnetic means.* It is a short step to the conclusion that it is also possible to induce *unstable* thermonuclear reactions by a similar method.

This led us to a bold conjecture on the possible weaponization of such a phenomenon.

- Every object or material substance has a resonant frequency.
- Sound waves that are resonant to that frequency can be amplified.
- Once amplified, they can be projected and accelerated to that object by electromagnetic forces – rather like a radio wave carries, or is "modulated" by acoustic information.
- Once the carrier electromagnetic wave reaches the target, the acoustic energy enters inside of the target, resonating harmonically with it, and loading energy into it.
- The threshold of instability is reached and the target is disrupted at the sub-atomic level by breaking down the bond between the sub-atomic particles.
- It is most effective if the carrier beam as well as its acoustic modulation is itself resonant with the target.

If one now recalls all that has been said thus far about phase conjugate mirrors, the coupling of non-local systems and the various forms of energy – gravitational, acoustic, electromagnetic – as well as time-reversed phase conjugation, *and engineers these components together,* one comes very close to an understanding of the awesome weapon the Great Pyramid was, and how it worked, for as we shall discover in chapter five, not only did it have an electromagnetic cohered output in the form of a maser, but also very likely an acoustic amplification and modulation of that output. One may speak of it as being a phase-conjugate electro-acoustic "howitzer."

Selecting the region is simple: one merely configures the antennae to be sufficiently out of phase to guide the interference signal to the point above the planet's surface one wishes to "heat". Converging there, the signals supercharge the ionosphere, causing the charge differential, which would eventually be discharged to the surface in several massive bolts of electricity per second, each bolt *far exceeding anything in the largest natural thunderstorm.* A surface target would be obliterated in an electrical blast much like a large H-bomb without the collateral radioactivity of such bombs.

In short, one looks for evidence of a unified physics and technology in the Great Pyramid that would permit it to be used or deployed as a weapons system that would have permitted a variety of specific applications:

- Weather modification
- Communications disruption (or enhancement)
- Defensive shielding
- Offensive use for mass destruction

With respect to each of these, I believe sufficient paleographical evidence has been presented in chapter two, particularly in Sitchin's texts, to document all four. Certainly the oldest religious traditions all bear witness to a catastrophic inundation of the world that was due to the wickedness of humanity.

So what remains is to discover whether or not the Pyramid *coupled* such energies together via the principle of a coupled harmonic oscillator. If so, then one should expect to encounter, at some point or points in its design, the following features:

- A coupling to inertial or gravitational phenomena, such as the wobbling of the earth via the precession of the equinoxes, the center of galactic mass, and so on.
- A coupling to electromagnetic phenomena, such as the mean temperature of the earth, the velocity of light, and so on.
- A coupling to the fundamental mathematical and physical constants.
- A coupling to fundamental acoustical energy, such as the Schumann resonance of the earth, and so on.
- Evidence of a coupling to, or use of, cohered electromagnetic energy (lasers, masers, etc).
- Evidence of a coupling to, or use of, nuclear energy.

159

- Evidence of coupling to the primary differential of paleophysics, time, in such a way as to indicate a fundamental reference time, or "base time" or "time lock.

These things must be born in mind as we now take a quick tour and survey, of the structure itself.

V.
A Quick Tour

"It is reasonable to assume that if we were to destroy ourselves through nuclear holocaust, the geological and biological record would bear witness to it and reveal that knowledge to future archaeologists as they became more advanced in their science. At the same time, some of our civil engineering projects might survive, and the occasional archaeological anomaly might turn up to promote some thought in that direction."
Christopher Dunn, <u>The Giza Power Plant</u>[1]

A. A Physical Tour

An aerial view of the Giza compound affords the best place to begin our tour.

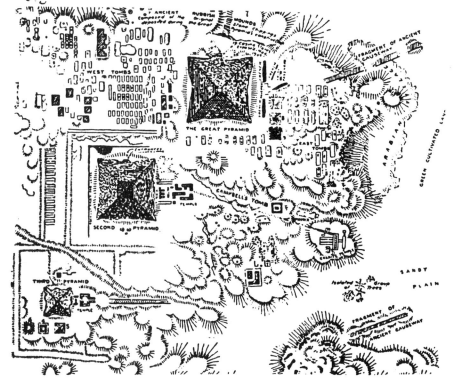

[1] Christopher Dunn, *The Giza Powerplant*, p. 244.

To the top right, one sees the Great Pyramid, its parabolic faces clearly in evidence. Notice the fragment of what most archaeologists call a "causeway" leading away to the east-northeast. Note that if one extends the causeway to the Great Pyramid itself, it appears to intersect at the center or just to the south of the center of the eastern parabolic face.

Below and to the left of the Great Pyramid we find the second Pyramid, the other dominant structure at Giza. Notably, in Piazzi Smith's drawing which is reproduced here, it too appears to have slightly parabolic faces, though not as pronounced as the Great Pyramid. This may or may not be an inaccuracy in the drawing; nevertheless, it is good to point it out, as some researcher may undertake to survey that pyramid as accurately as the Great pyramid. Notice that to the right of the Second Pyramid one finds a "temple", with the traces of yet another "causeway" extending to the east southeast past the Sphinx to a granite "temple."

Finally, in the lower left-hand corner of the diagram, one finds the Third Pyramid, the smallest of the three large pyramids at Giza. Again, one notes the peculiar feature in Piazzi Smith's drawing that it too appears to have "parabolic" faces, and, like the other two pyramids, a "causeway" leading almost due east. Immediately south of the Third Pyramid are the fourth, fifth, and sixth pyramids, structures of evidently inferior construction when compared with the first three.

Now let us note one feature unique to the Great Pyramid. In thousands of years, this massive building has settled less than half an inch, in spite of numerous earthquakes in the region. Since Petrie's comprehensive survey of Giza, engineers have known why. Beneath the Great Pyramid there are five massive stones or "sockets", four at each corner of the structure, and a fifth on the diagonal above the southeast corner (Cf. Figure Two). These sockets are a "ball and socket" joint familiar to modern engineering, permitting the building to rock and shift gently when the earth moves. This is the surest evidence, in and of itself, that the Great Pyramid is a coupled oscillator, for this feature is analogous to pressing down a key of a piano silently while striking another key to make it resonate freely. The Pyramid, in short, was *designed to move.*

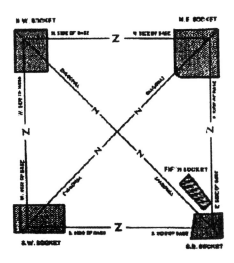

Figure Two:
The Five Sockets of the Base of the Great Pyramid

If one now looks at a North-South cross section of the Great Pyramid, one is immediately confronted by an anomaly, as far as the pyramids at Giza go, for alone of all the pyramids there, the Great Pyramid has internal chambers *above* the ground line in addition to a subterranean chamber (Figure Three). To the left of the axis running from the apex of the Pyramid to its base, one finds a large chamber, capped by five layers of huge roughly hewn granite stones beneath a corbeled roof. This is the so-called "King's Chamber". From this chamber two thin shafts emerge diagonally upward to the north and south faces of the structure. These are called "air shafts." Immediately to the right of the King's Chamber is a smaller chamber, called the "Antechamber", and then, extending diagonally downward, a tall and narrow "Grand Gallery", which ends in a shaft intersecting with another shaft leading below ground to the "Subterranean Chamber."

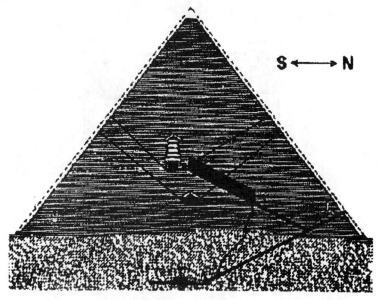

Figure Two:
A North-South Cross Section of the Great Pyramid

At the lowest point of the Grand Gallery, a straight passageway extends to the lower chamber, the "Queen's Chamber", a lower and less massive chamber than the King's Chamber, but which likewise has a corbeled roof. Notice that the apex of the Queen's Chamber roof and the upper end of the Grand Gallery – the so-called "Great Step" - all lie on the axis running through the center of the structure up to the apex of the Pyramid itself. From the Queen's chamber two more "air shafts" make their way diagonally upward to the north and south faces of the Pyramid, but do *not* actually emerge on the surface of the faces, but stopping just short of it. Yet another anomalous feature. Note also the layers or stone courses of the Pyramid, an important feature that we will discuss in more detail in chapter six.

Now let us look a little closer at each of these chambers, beginning with the Grand Gallery. In terms of sheer size, this is the largest of the interior chambers of the Great Pyramid, and it has a number of unusual features (cf. Figure Three). Along each side of the Gallery, there is a narrow flat section, into which twenty-seven notches are cut at equal distances on each side. Not only this, but the Grand Gallery's walls narrow from the bottom to the top, and the stones of its roof are sloped.

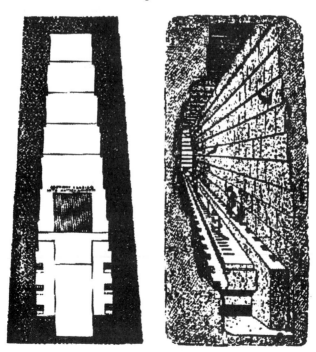

Figure Three:
The Grand Gallery: Cross Section and Perspective Views

At the end of the Grand Gallery there is a "low passage" leading to the "antechamber", followed by another low passage leading into the King's Chamber (Figure Four). Inside the King's Chamber one finds a large oblong granite box, one corner of which looks as if it has been melted, called the Coffer. Note the "air shafts" leading up from the King's Chamber.

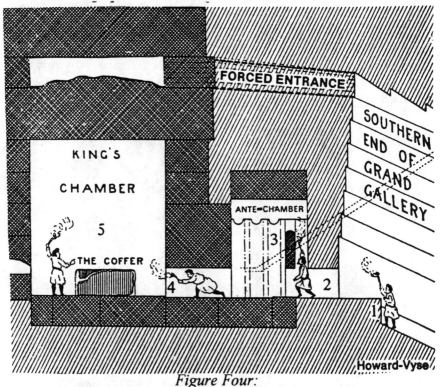

Figure Four:
The End of the Grand Gallery, The Antechamber, King's Chamber, and Coffer.

Before we enter the King's Chamber, there are a number of peculiar things we must observe in the Antechamber. First, look at Haberman's perspective view of the Antechamber (Figure Five).

Notice that immediately after you emerge from the first low passage, there is a large slab with what appears to be a protrudance resembling a horseshoe in the center (Figure Six). If one were to remove this slab, one could see the rest of the Antechamber (Figure Seven). Looking again at Figures Four and Five, you will also note, on each side of the Antechamber behind the slab, three large slots, at the top of which are three semi-circular notches. This feature has led some to speculate that at one time there were three movable slabs of rock set into these notches which could be raised or lowered rather like a portcullis. As we shall see, Christopher Dunn has a most ingenious explanation of what purpose these notches and the portcullis may once have served.

Continuing our journey into the King's Chamber itself, we find the Coffer with its "melted" corner lying toward the western wall of the Chamber (cf. Figures Eight and Nine). Looking at the north-south cross section of the King's Chamber (Figure Nine), we discover five layers of very large granite slabs, with flat bottoms but very roughly carved tops, yet another seeming anomaly in a structure so perfectly constructed. Looking for a moment longer, we discover in the three uppermost chambers the only hieroglyphics ever found inside the Great Pyramid, and found in a most unlikely place, namely, not in the chamber that is supposed to contain the actual sarcophagus, the Coffer.[2] If one looks at the east-west cross section (Figure Eight), one notes that the large granite stones comprising the roofs of the five chambers above the King's Chamber are all vertically cantilevered. One notes also the passageway that Vyse forced up through the stone courses to reach these chambers. Once on the inside, he reported that he became covered with a fine powdery black soot. Finally, note where the low passageway emerges in the King's Chamber. Just to the left

---

[2] The British archaeologist Howard Vyse claimed to have "discovered" these in the nineteenth century, much to the relief of orthodox Egyptology. The problem was, the hieroglyphics themselves are of a very illiterate nature, not what one would expect from such a lavish expenditure of effort for a tomb. This fact has led many to speculate that Vyse in fact forged the hieroglyphs himself in order to lay claim to a significant discovery. I adhere to this view.

one will see the entrance of the "air shaft" into the chamber, and on the south wall directly opposite, the other shaft emerges into the chamber.

Al Mamun in the Grand Gallery.

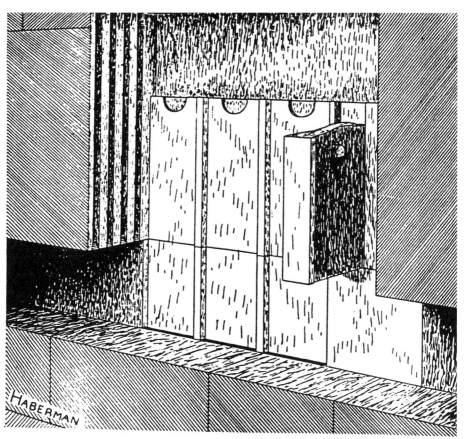

Figure Five: Haberman's Perspective View of the Antechamber

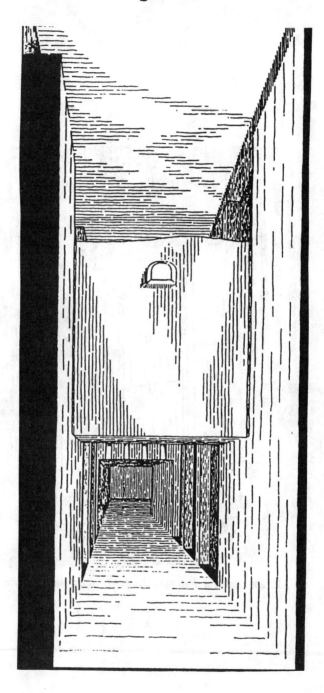

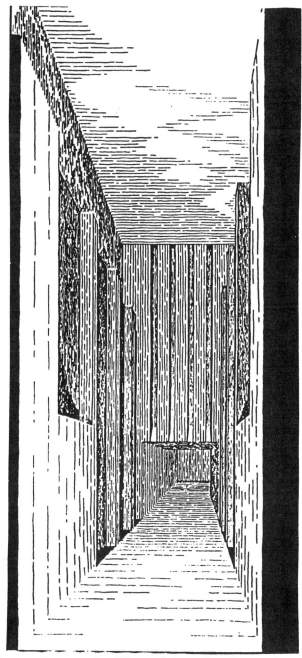

Figure Seven: Perspective View of Antechamber without Slab

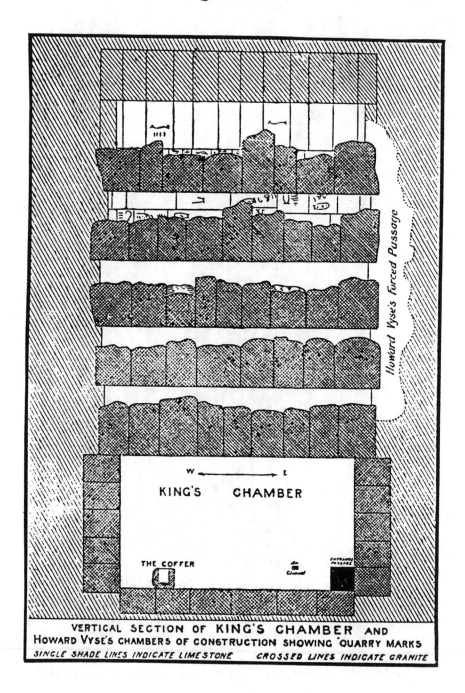

VERTICAL SECTION OF **KING'S CHAMBER** AND
HOWARD VYSE'S CHAMBERS OF CONSTRUCTION SHOWING QUARRY MARKS
SINGLE SHADE LINES INDICATE LIMESTONE   CROSSED LINES INDICATE GRANITE

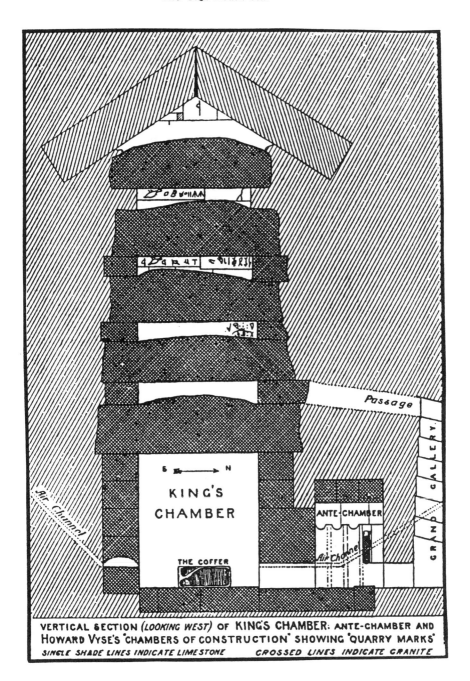

VERTICAL SECTION *(LOOKING WEST)* OF KING'S CHAMBER: ANTE-CHAMBER AND HOWARD VYSE'S 'CHAMBERS OF CONSTRUCTION' SHOWING 'QUARRY MARKS'

SINGLE SHADE LINES INDICATE LIMESTONE   *CROSSED LINES INDICATE GRANITE*

B. A Mathematical and Physical Survey

(1) Universal Geometric, Mathematical, and Physical Properties

Now let us leave aside, until the next chapter, our physical journey, and note some of the more prominent mathematical and physical facts about the Great Pyramid. The first, and perhaps most unusual, feature of the structure is that it appears to embody a system of measure very much like our English system of measurement. The so-called "pyramid inch" or PI is equal to 1.0011 of our present American inches and 1.0010846752 British inches.[3]

Equally as strange is the man who is credited with this discovery. It was Sir Isaac Newton who found that "many of the measurements of the pyramid would be in whole numbers if this unit (or measure) were used."[4] It becomes much more anomalous. For example, if one measures the perimeter of the base of the Pyramid and divides it by twice the height, one will get a result approximating the value of π to five decimal places.[5] This is significant, for it means that long before the Greeks "discovered" π, the paleoancient builders of the Pyramid already knew about it. Perhaps the Greeks merely rediscovered it, or perhaps were allowed to publicize it.

The high strangeness of it all only accumulates. For example, if one computes the ratio of the apothem (the face slant to height) to half the length of a base of the Pyramid, one obtains yet another mathematical constant, that of ϕ, which has a theoretical value of 1.61818.... ϕ, as mathematicians and physicists know, has some rather remarkable properties of its own. If one adds 1 to ϕ, one will discover that $1+\phi = \phi^2$. Likewise, $1+1/\phi=\phi$. ϕ is also the basis for

---

[3] Rodolfo Benavides, *Dramatic Prophecies in the Great Pyramid* (1974), p. 2. The best compilation of the mathematical and physical properties is that of Mr. Tim G. Hunkler, at <www.hunkler.com> from which this information is gathered.

[4] John Zajac, *The Delicate Balance* (1989), p. 151.

[5] Benavides, op. cit., p. 24.

the Fibonacci sequence 1,1,2,3,5,8,13,21,43 and so on, which was not a mathematical fact well known until ca. 1200 AD![6]   There even exists evidence that the Pyramid's builders "worked out a relationship between π and ϕ of π = 6/5ths of ϕ^2."[7]   Strange indeed, especially if all this is simply to entomb a dead king.

But it doesn't stop there. The Pythagorean relationship represented by a 3-4-5 right triangle can be found in the dimensions of the King's Chamber.   The east wall's diagonal is 309", its length is 412", and its long central diagonal is 515".   Here again, whoever built the Great Pyramid anticipated the Greeks by several thousands of years, for the Pythagoreans are not credited with the discovery until about 495 BC.[8]

The most extraordinary mathematical and physical constant of them all, however, is the fact that Gaussian constant of gravitation (k) is expressed trigonometrically as "the reciprocal of the distance between the Coffer and the north or south wall of the King's Chamber, minus one ten-billionth the bottom perimeter of the Coffer.... (or) r degrees, 5 minutes, 49.96 seconds of arc."[9]   The constant of gravitation, to bury a Pharaoh?

Absurd!

(2) Celestial Alignments and Properties

Robert Bauval and Adrian Gilbert state that "the pyramid was built circa 2450 BC according to star alignment data for the 4 air shafts of the King (sic.) and Queen's chambers."[10]   But the conclusion exceeds the evidence. Given the extraordinary degree of mathematical, physical, and astronomical data evident in the Pyramid's dimensions, it is entirely possible that it was built much earlier with the knowledge that those shafts would align at that

---

[6] J.P. Lepre, *The Egyptian Pyramids: A Comprehensive and Illustrated Reference* (1990), p. 126.

[7] Ibid., p. 194.

[8] Ibid., p. 102.

[9] Gray, op. cit., p. 275.

[10] Robert Bauvall and Adrian Gilbert, *The Orion Mystery* (1994), p. 35.

moment. We simply do not know. The point is that alignments at certain periods with certain features of the Pyramid are an insufficient basis to date its building. However, we may point out that such a celestial and zodiacal alignment *does* link the Pyramid with a constellation subsequent astrological lore associates with death. In that sense, too, it is a death star, for it is aligned with the death star, Orion.

The length of a base side of the Pyramid is 9131 pyramid inches. But expressed in yet another peculiar "pyramid" unit of measure, the "pyramid cubit", the length is exactly 365.24 cubits, exactly the number of days in a terrestrial year.[11] Each base side of the Pyramid, in other words, is an exact temporal measure coupling to the time it takes for the earth to complete one revolution in its orbit around the sun. This isn't all, for it was also built to indicate exactly when the earth would pass through its solstices and equinoxes,[12] and, at exactly noon during the spring equinox, due to the precise angles of its faces, it casts no shadow whatsoever.[13]

It is also built in a ratio of the mean distance of the earth to the sun, for its height times 10^9 equals the mean radius of the earth's orbit of the earth around the sun, the basic astronomical unit.[14] Likewise, if one doubles the perimeter of the bottom of the Coffer, and multiplies it by 10^8, one obtains the mean distance to the moon.[15] One may also find the ratio of the sun's radius expressed as a function of the measure of the perimeter of the Coffer,[16] as well as another peculiar ratio. "The pyramid embodies a scale ratio of 1/43200. The height times 43,200 = 3938.685 miles, which is the polar radius of the earth within 11 miles."[17]

---

[11] Gray, op. cit., p. 5.

[12] Benavides, op. cit., p. 9.

[13] Gray, op. cit., p. 111.

[14] Benavides, op. cit., p. 11.

[15] Gray, op. cit., p. 106.

[16] Ibid., p. 267.

[17] Graham Hancock and Robert Bauvall, *The Message of the Sphinx* (1998), p. 38.

The most astonishing astronomical feature embodied in it is the precession of the equinoxes. If one measures the distance from the ceiling of the King's Chamber to the apex of the Pyramid, one will get 4110.5 pyramid inches. This is the radius of a circle whose circumference yields a numerical value giving the number of years it takes to complete a precession of the equinox: 4110.5 x 2 x π = 25,827.[18]

Not surprisingly, one also finds an accurate measure of the velocity of light itself, for is the distance of one astronomical unit is known – and we have seen its builders knew it – and the transit time of light for this distance is also known, then the velocity of light can be calculated.

So when *does* one date the structure? Carbon-14 dating of the Great Pyramid's mortar point to a date of construction ca. 2800 BC. However, the celestial alignments of "the pyramid positions on the ground are a reflection of the positions of the stars (in the belt of) the constellation Orion circa 10,400 BC."[19] More recently, it was discovered that radioactive dating of the stones at the top of the Pyramid indicated that they were older than the stones at its base!

But this curious feature can be explained *if one presupposes that at one time nuclear reactions took place inside of it. The exposure curve of radiation for the stones would, in that case, make sense.* Moreover, exposure to such intense radiation would in fact distort carbon-14 dating of the Pyramid's mortar, making it appear far younger than it in fact may be.

---

[18] Benavides, op. cit., p. 22.

[19] Bauvall and Gilbert, op. cit., p. 124.

(3) Terrestrial Properties and Alignments

The mean density of the earth is approximately 5.7 times that of water at 68 degrees Fahrenheit at a barometric pressure of 30 pounds per inch. In the King's Chamber, all of the stone courses have 23 or more stones except for the 5th course, which contains only 7. Thus, encoded in the fifth course of the King's chamber is the mean density of the earth.[20]

Moreover, "there is so much stone mass in the pyramid that the interior temperature is constant and equals the average temperature of the earth, 68 degrees Fahrenheit."[21] In other words, the mean thermal gradient of the pyramid is exactly that of the earth. And since the mean thermal gradient of the earth is the result of a constellation of factors such as distance from the sun, the amount of radioactivity absorbed from the sun, its orbital velocity, mean density, electromagnetic field strength, the rotational tilt of its axis and so on, the pyramid thus reflects an extremely accurate knowledge of solar and terrestrial physics.

The Great Pyramid is also the most accurately aligned building in the world. It is aligned to true north with only 3/60[th]s a degree of error. Likewise, it is located at the exact center of the surface of the land mass of the earth, since the east-west parallel and the north-south meridian that both cross the most land intersect at only two places on earth, one in the ocean and another precisely at the Great Pyramid.[22] That's not all. The average height of land above sea level is approximately 5449 inches, which is also the Pyramid's height.[23] Stop and consider what this means. Not only did the civilization that built the Pyramid have to possess advanced topographical data of the entire surface of the earth, it would have also have had to possess extremely sophisticated mathematical

---

[20] Julian T. Gray, *The Authorship and Message of the Great Pyramid* (1953), p. 255.

[21] Benavides, op. cit., p. 40.

[22] Ibid., pp. 71-72.

[23] Zajac, op. cit., p. 153.

techniques in order to calculate such a measurement accurately. Additionally, in order to embody all these features in one structure, it would have had to possess computer-aided design and architectural technology of some sort analogous to our own.

Celestial, solar, lunar, terrestrial alignments accurately reproduced over and over again. And we have touched on but a very few of a vast inventory. Strange, if not downright weird, construction features. Accurate ratios of the thermal and mass gradients of the earth, the astronomical unit, the precession of the equinoxes, the average height of land above sea level. And all this to bury a pharaoh? Surely not, says Christopher Dunn. It was not a tomb.

It was a machine.

VI.
The Machine Hypothesis

"A credible theory would have to explain..."
Christopher Dunn, <u>The Giza Power Plant</u>

A. A Credible Theory

Author and engineer Christopher Dunn proposes the most lucid account of the machine hypothesis in his book *The Giza Power Plant*. This crucial book in the growing literature on the Great Pyramid can only be summarized here, but every effort will be made to cite Dunn's actual words. It cannot be too strongly emphasized that this is only a summary of his work, and no summary can substitute for a careful study of his illuminating and important analysis.

For Dunn, any credible theory about the Great Pyramid would have to account for the following anomalous facts:

- The selection of granite as the building material for the King's Chamber. It is evident that in choosing granite, the builders took upon themselves an extremely difficult task.
- The presence of four superfluous chambers above the King's Chamber.
- The characteristics of the giant granite monoliths that were used to separate these so-called "construction chambers."
- The presence of exuviae, or the cast-off shells of insects, that coated the chamber above the King's Chamber, turning those who entered black.
- The violent disturbance in the King's Chamber that expanded its walls and cracked the beams in its ceiling but left the rest of the Great Pyramid seemingly undisturbed.
- The fact that the guardians were able to detect the disturbance inside the King's Chamber, when there was little or no exterior evidence of it.
- The reason the guardians thought it necessary to smear the cracks in the ceiling of the King's Chamber with cement.
- The fact that two shafts connect the King's Chamber to the outside.

180

- The design logic for these two shafts – their function, dimensions, features, and so forth.

Any theory offered for serious consideration concerning the Great Pyramid also would have to provide logical reasons for all the anomalies we have already discussed and several we soon will examine, including:

- The antechamber
- The Grand Gallery, with its corbeled walls and steep incline.
- The Ascending Passage, with its enigmatic granite barriers.
- The Well Shaft down to the Subterranean Pit.
- The salt encrustations on the walls of the Queen's Chamber.
- The rough, unfinished floor inside the Queen's Chamber.
- The corbeled niche cut into the east wall of the Queen's Chamber.
- The shafts that originally were not fully connect to the Queen's Chamber.
- The copper fittings discovered by Rudolph Gantenbrink in 1993.
- The green stone ball, grapnel hook, and cedar-like wood found in the Queens' Chamber shafts.
- The plaster of paris that oozed out of the joints made inside the shafts.
- The repugnant odor that assailed early explorers.[1]

While Dunn's work admirably fulfills these criteria, there are nevertheless some omissions from this list that must be accounted for as well. What was all this engineering designed to do? Dunn's answer is obvious from the title of his work: it was to produce power. But power for *what*?

As has been demonstrated, there are several facts maintained in ancient texts and traditions that are immediately relevant to a proper understanding of the machine hypothesis, all of them more or less answering "What *kind* of machine was it?" Any credible theory must therefore also explain the following additional factors, explored in chapters two, three, and four, in addition to Mr. Dunn's list:

---

[1] Christopher Dunn, *The Giza Power Plant: Technologies of Ancient Egypt* (Santa Fe, New Mexico: Bear and Company Publishing, 1998), pp. 46-47.

- The consistent and pervasive religious tradition in Egypt associating the compound with Zodiacal symbols associated with death and immortality.
- The ancient texts, traditions, and archaeological sites consistent with the use of advanced weapons systems of mass destruction that were mentioned in chapters 2, 3, and 4.
- The ancient texts and traditions that indicated the one-time existence of a very advanced theoretical physics.
- The ancient texts that indicate the Great Pyramid was a weapon.
- The possible relationship of the Great Pyramid to the other structures of Giza in the performance of that function.[2]

Since so much of the weapons hypothesis is related to Dunn's work, the basic functions of the physical features of the Great Pyramid in his model will be summarized here before exploring how they may or may not relate to the weapon hypothesis in chapter six. Dunn's hypothesis will be summarized here in more or less the order he himself develops it.

B. Extremely Close Tolerances
and Some Provocative Questions

Engineers who have studied the Great Pyramid consistently come away amazed, and utterly baffled by its unusually close tolerances. Indeed, notes Dunn, it was this factor that compelled his interest in the structure.

> Here was a prehistoric monument that was constructed with such precision that you could not find a comparable modern building. More remarkable to me was that the builders evidently found it *necessary* to

---

[2] As we shall discover, Dunn alone, of all the literature on the subject, proposes a purely functional purpose to some of the other pyramids at Giza.

maintain a standard of precision that can be found today in machine shops, but certainly not on building sites.[3]

But why were such tolerances necessary for a building that was designed principally as a religious tomb and/or astronomical observatory? Why have such tolerances at all? And how were they achieved?[4] Dunn's answer does not require the reader to subscribe to the dubious notion that the whole structure was built to such tolerances to ensure the Pharaoh's immortality.

> I consider two possible alternative answers. First, the building was for some reason *required* to conform to precise specifications regarding its dimensions, geometric proportions, and its mass. As with a modern optician's product, any variation from these specifications *would severely diminish its primary function.* In order to comply with these specifications, therefore, greater care than usual was taken in manufacturing and constructing this edifice. Second, the builders of the Great Pyramid were highly evolved in their building skills and possessed greatly advanced instruments and tools. The accuracy of the pyramid was normal to them, and perhaps their tools were not capable of producing anything less than this superb accuracy, which has astounded many over the years. Consider, for example, that the modern machines that produce many of the components that support out civilization are so finely engineered that the most inferior piece they could turn out is more accurate than what was the norm for those produced one hundred years ago. In engineering, the state of the art inevitably moves forward.[5]

These two provocative observations require some comment, beginning with Dunn's idea that the extreme tolerances may just be the coincidental effect of a society possessed of a highly developed skill in engineering.

And that is the point. If such skills exceed the capacity of our own most advanced construction capabilities, then one is dealing with a civilization more advanced than our own. And as will be seen in chapter six, the most pervasive employment of such

---

[3] Dunn, op. cit. p. 51.

[4] Dunn, op. cit., p. 56.

[5] Ibid., p. 64, emphasis added.

construction tolerances in our own society is often in connection with military projects or extremely advanced optics, often both.

The second point is this: if these extremely close tolerances were necessary to the proper functioning of the structure, then one is faced with something of an anomaly for which no contemporary analogy actually exists, and so one is left to make educated conjectures. Our own contemporary notions of power and energy would not seem to require such close tolerances for the construction of a mere power plant, *unless* the paleoancient notion of power and energy was fundamentally different than our own, based on a kind of unified field physics that was *practical and testable*, a goal that we have not yet achieved. Indeed, Dunn never satisfactorily explicitly explains the necessity for such close tolerances in a mere "power plant." However, this is not a deficit to his work, as he is concerned not so much with conjecture, but merely the evidence that the Pyramid was a type of machine involving enormous power output.

C. Advanced Machining and Ultra-Sonic Drilling

One of the most provocative and most thorough considerations of the advanced technology used in the Great Pyramid is Dunn's discussion of the evidence of advanced machining in the building.[6] In this respect, his analysis of the Coffer in the King's Chamber is the most anomalous of all the evidence for a sophisticated technology exceeding our own contemporary abilities.

> Along with the evidence on the outside of the King's Chamber coffer, we find further evidence of the use of high-speed machine tools on the inside of the granite coffer. The methods that were evidently used by the pyramid builders to hollow out the inside of the granite coffers are similar to the methods that would be used to machine out the inside of components today. Tool marks on the coffer's inside indicate that when the granite was hollow out, workers made preliminary roughing cuts by drilling holes into the granite around the area that was to be removed.[7]

---

[6] Dunn, op. cit., pp. 67-91.

[7] Ibid., pp. 79-80.

*The Coffer, in
the King's Chamber,*

Elevation, looking West.

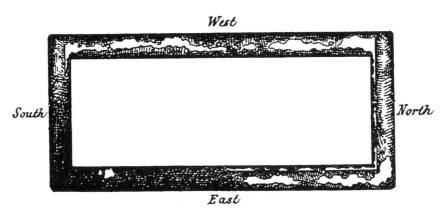

*Plan, looking from above;
the shading in proportion to the deviation
from a horizontal plane.*

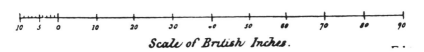

Scale of British Inches.

185

Dunn then reproduces the following figure:

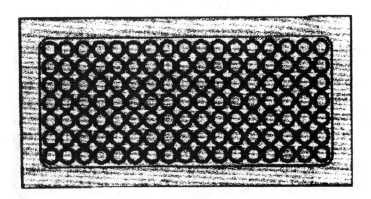

The fact that the inside of the Coffer appears to have been drilled was one of the most anomalous facts noted by the famous nineteenth century pyramidologist, Sir William Flinders Petrie.

> To an engineer in the 1880s, what Petrie was looking at was an anomaly. The characteristics of the holes, the cores that came out of them, and the tools marks would be an impossibility according to any conventional theory of ancient Egyptian craftsmanship, even with the technology available in Petrie's day. Three distinct characteristics of the hole and core...make the artifacts extremely remarkable:
>
> - A taper on both the hole and the core.
> - A symmetrical helical groove following these tapers showing that the drill advanced into the granite at a feedrate of .1o inch per revolution of the drill.
> - The confounding fact that the spiral groove cut deeper through the quartz than through the softer feldspar.[8]

But in 100 years, as technology has advanced, the anomaly has deepened:

> In conventional machining the reverse would be the case. In 1983 Donald Rahn of Rahn Granite Surface plate Co. told me that diamond drills, rotating at nine hundred revolutions per minute, penetrate granite

---

[8] Dunn, op. cit., p. 84.

at the rate f one inch in five minutes. In 1996, Eric Leither of Tru-Stone Corp. told me that these parameters have not changed since then. The feedrate of modern drills, therefore, calculates to be .0002 inch per revolution, indicating that the ancient Egyptians drilled into granite with a feedrate that was five hundred times greater or deeper per revolution of the drill than modern drills! The other characteristics of the artifacts also pose a problem for modern drills. Somehow the Egyptians made a tapered hole with a spiral groove that was cut deeper through the harder constituent of the granite. If conventional machining methods cannot answer just one of these questions, how do we answer all three?[9]

Dunn's answer explains the advanced drilling technique used to hollow out the Coffer, but in the process, only validates the existence of an extremely sophisticated technology in paleoancient times.

In contrast, ultrasonic drilling fully explains how the holes and cores found in Valley Temple at Giza could have been cut, and it is capable of creating all the details that Petrie and I puzzled over. Unfortunately for Petrie, ultrasonic drilling was unknown at the time he made his studies, so it is not surprising that he could not find satisfactory answers to his queries. In my opinion, the application of ultrasonic machining is the only method that completely satisfies logic, from a technical viewpoint.

Ultrasonic machining is the oscillatory motion of a tool that chips away material, like a jackhammer chipping away at a piece of concrete pavement, except much faster and not as measurable in its reciprocation. The ultrasonic tool bit, vibrating at 19,000- to 25,000-cycles-per-second (hertz), has found unique application in the precision machining of odd-shaped holes in hard, brittle material such as hardened steels, carbides, ceramics, and semiconductors. An abrasive slurry or paste is used to accelerate the cutting action.

The most significant detail of the drilled holes and cores studied by Petrie was that the groove was cut deeper through the quartz than through the feldspar. Quartz crystals are employed in the production of ultrasonic sound and, conversely, are responsive to the influence of vibration in the ultrasonic ranges and can be induced to vibrate at high frequency. When machining granite using ultrasonics, the harder material (quartz) would not necessarily offer more resistance, as it

---

[9] Dunn, op. cit., p. 84.

would during conventional machining practices. An ultrasonically vibrating tool bit would find numerous sympathetic partners, while cutting through granite, embedded right in the granite itself. Instead of resisting the cutting action, the quartz would be induced to respond and vibrate in sympathy with the high-frequency waves and amplify the abrasive action as the tool cut through it.[10]

An amazing anomaly indeed, for if ultrasonic drilling is a machining technique found only at the end of the twentieth century, then this would seem to imply that the palaeoancient Very High Civilization achieved *at least* a similar level of technological and scientific sophistication to our own.

Finally, there are two other facts about the Coffer that must be mentioned. First, the Coffer is one solid block of granite that has been hollowed out, probably either with ultrasonic drilling or with some technique as yet still unknown to us. And this raises a question: why, if the Coffer was meant to be a sarcophagus, why was it necessary for its builders to go through the extra complication of machining it in this fashion? Why not build it out of several pieces, as the Egyptians were known to do in other instances?[11] Second, the Coffer is a precisely machined object, not showing the slightest imperfection, *which means that it was constructed inside the King's Chamber.* Its builders intended for it to be precise for some as yet unknown reason.

> They had gone to the trouble to take the unfinished product into the tunnel and finish it underground for a good reason. It is the logical thing to do if you require a high degree of precision in the piece that you (Sic.) are working. To finish it with such precision at a site that maintained a different atmosphere and a different temperature, such as in the open under the hot sun, would mean that when it was finally installed in the cool, cavelike temperatures of the tunnel, the workpiece would lose precision. The solution then as now, of course, was to prepare precision objects in a location that had the same heat and humidity in which they were going to be housed.[12]

---

[10] Dunn, op. cit., p. 87.

[11] Ibid., p. 95.

[12] Ibid., p. 97.

As has been seen, the Great Pyramid's builders so constructed the Pyramid that the King's Chamber maintains a temperature that is very close to the mean thermal gradient of the earth.

D. How it Worked

Dunn's analysis of the chambers and passages of the Great Pyramid and their possible functions simply cannot be bested. It is the most comprehensive survey of their potentialities based upon known science and currently existing technology, and therefore what follows is but a crude summary of Dunn's excellent work. It is crucial to the basis of our speculations in the next chapter, so some detailed understanding of Dunn's hypothesis is essential.

1. Missing Components and Many Possible Solutions

In addition to the advanced machining that so mystified Petrie, Dunn points out that the design of the inner chambers and passageways of the Great Pyramid seem to connote some purely functional purpose having little to do with the death-resurrection-Osiris mythology of ancient Egypt. "I became convinced that I was looking at the prints for an extremely large machine, *except this machine had been relieved of its inner components for some inexplicable reason.*"[13] This remark is truly astonishing, for nowhere in Dunn's work is any reference made to the ancient texts cited by Zechariah Sitchin that indicate that components were indeed removed from the Great Pyramid – some to be forever destroyed – by the victors in the "Second Pyramid War." Indeed, nowhere does Dunn refer to Sitchin's work at all. His approach, as an engineer simply examining the evidence the Pyramid presents, is to extrapolate from that evidence and known engineering and scientific principles the possible function of the Pyramid. On that basis, he concluded, "something is missing," corroborating apparently independently the ancient texts cited by Sitchin.

---

[13] Dunn, op. cit., p. 122, emphasis added.

189

But what kind of machine? Dunn maintains an open mind: "In proposing my theory that the Great Pyramid is a power plant, I am not adamantly adhering to any one proposition. The possibilities may be numerous. However, the main facts are inescapable, for they were noted many years ago, and it would be impossible for an open-minded, logically thinking person to disregard them."[14] If Dunn does not go all the way to the weapon hypothesis, he *does* allude to a potentially destructive use of the technology behind the Great Pyramid if not of the structure itself.[15]

2. Some Elementary Physics: Coupled Harmonic Oscillators and Damping

The principle of a coupled harmonic oscillator in resonance to some fundamental "can unleash an awesome and destructive power."[16] The earth, as any college general science textbook will explain, is both a source of tremendous mechanical energy as well as of electro-magnetic energy, witness the enormous power unleashed in an earthquake or a thunderstorm. Normally, mechanical and electromagnetic energy propagates in two kinds of waves, transverse "S" waves and longitudinal "P" waves. "Primary or compressional waves (P waves) send particles oscillating back and forth in the same direction[17] as the waves are travelling. Secondary or transverse shear waves (S Waves) oscillate perpendicular to their direction of travel. P waves always travel at higher velocities than S waves and are the first to be recorded by a seismograph."[18]

---

[14] Dunn, op. cit., p. 123.

[15] Ibid., pp. 243-245.

[16] Ibid., p. 136. It should go without saying that the basic principle of resonance, as stated here, implies a potential for weaponization.

[17] The term "direction" may be misleading to some. What Dunn meant to say is "the same *axis*".

[18] Dunn, op. cit., p. 126. The "wave-particle" duality of current quantum mechanics and light theory is well known, but perhaps the "duality" is itself not properly understood, or at least, referred to in the scientific literature. Perhaps the duality is best *expressed* as the transverse-longitudinal wave duality. This

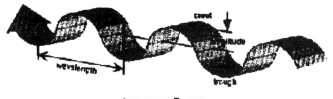

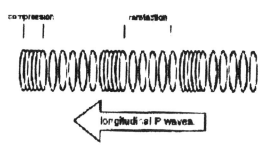

Transverse (S) Wave and Longitudinal (P) Wave

This relationship between mechanical, or acoustic types of waves, and electromagnetic waves, is deeply mysterious, but yet, is a commonplace that most people are familiar with. Dunn puts it this way:

> Turn on any motor or generator and you can hear the energy at work: the motor/generator will hum as it revolves. This hum is associated with the energy itself and not so much the movements of the rotor through the air. This phenomenon is evident when a motor stalls when the power is still turned on. When too great a load is put on a motor, and the motor stalls, the hum will become louder. The electrical and magnetic forces in the motor generate the sound waves. The earth itself, as a giant dynamo, produces similar sound waves.... Collectively

form of stating the duality would thus seem to issue in a paradox: a photon of light would arrive at an observer before its detection or measurement by that observer. This would seem to suggest that recent experiments in "superluminal" phenomena might be being improperly interpreted: electromagnetic phenomena are "superluminal" but the observable phenomena are luminal and their wave characteristics are dependent upon the geometric configurations of the total system.

191

known as an electromagnetic 'cavity,' the elements that make it up are the Earth, the ionosphere, the troposphere, and the magnetosphere. The fundamental frequency of the vibrations is calculated to be 7.83 hertz, with overlaying frequencies of 14, 2o, 26, 32, 37, and 43 hertz.... The Earth's energy includes mechanical, thermal, electrical, magnetic, nuclear, and chemical action, each a source for sound. It would follow, therefore, that the energy at work in the Earth would generate sound waves that would be related to the particular vibration of the energy creating it and the material through which it passes.[19]

3.   Piezo-Electric Effect

But why use granite, one of the most difficult materials to work with, in constructing the Pyramid? Very simple, says Dunn. Granite is composed of billions of tiny quartz crystals suspended in the surrounding rock. Thus, if one stresses the granite, by pulsing it, each tiny quartz crystal would produce electrical output. This effect is known in physics as the piezoelectric effect.[20]

Any electrical stimulation within the Earth of piezoelectrical materials – such as quartz – would generate sound waves above the range of human hearing. Materials undergoing stress within the earth can emit bursts of ultrasonic radiation. Materials undergoing plastic deformation emit a signal of lower amplitude than when the deformation is such as to produce cracks. Ball lightening has been speculated to be gas ionized by electricity from quartz-bearing rocks, such as granite, that is subject to stress.[21]

Dunn produces the following diagram to accompany this comment.

---

[19] Dunn, op. cit., pp. 127-129.

[20] Piezo, meaning "stone". It is curious that the electrogravitics researcher and physicist Thomas Townsend Brown, whose other interests were known to have included UFOs, and who was alleged to have taken part in the design of the Philadelphia experiment, spent much of his last research investigating the electrical, magnetic, and acoustic properties of rocks.

[21] Dunn, op. cit., p. 129

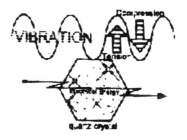

The Piezoelectric Effect

So the choice of granite is now rather obvious, for the weight of the Pyramid itself, pressing millions of tons of granite down through its stone courses, already places each tiny crystal under constant stress. Add to this the fact that the Pyramid's ball-and-socket construction allows it to *move* as a coupled harmonic oscillator means that all those quartz crystals are constantly being pulsed in resonance to the Schumann vibration of the earth itself. But note also, that this stress may also generate a cloud of ball lightening, an ionized plasma of gas that itself contains energy.

> When we question *why* there is a correlation between the earth's dimensions and the Great Pyramid, we come up with three logical alternatives. One is that the builders wished to demonstrate their knowledge of the dimensions of the planet. They felt it necessary to encapsulate this knowledge in an indestructible structure so that future generations, thousands of years in the future, would know of their presence in the world and their knowledge of it.[22]
>
> The second possible answer could be that the Earth affected the function of the Great Pyramid. By incorporating the same basic measurements in the pyramid that were found on the planet, the efficiency of the pyramid was improved and, in effect, it could be a harmonic integer of the planet.
>
> A third alternative may involve both the first and second answers. The dimensions incorporated in the Great Pyramid may have been included to demonstrate the builders' knowledge or more importantly,

---

[22] In other words, the "time capsule" hypothesis.

to symbolize the relationship between the Great Pyramid's true purpose and the Earth itself.[23]

For reasons discussed in the next chapter, I favor the second of these alternatives.

4. The Grand Gallery: An Acoustic Amplification Chamber and Helmholtz Resonators

Dunn is at his most brilliant when he analyzes the Grand Gallery and what its missing components may once have been. Noting that the ceiling tiles in the gallery tilt at an angle of approximately 45 degrees,[24] he observes that the Gallery is so constructed to be a massive acoustic amplification chamber, designed to amplify and reflect acoustic waves up the Gallery toward the Antechamber.

> The mystery of the twenty-seven pairs of slots in the side ramps is logically explained if we theorize that each pair of slots contained a resonator assembly and the slots served to lock these assemblies into place. The original design of the resonators will always be open to question; however, if their function was to efficiently respond (sic.) to the Earth's vibration, then we can surmise that they might be similar to a device we know of today that has a similar function – a Helmholtz resonator.[25]

A classic Helmholtz resonator is a hollow sphere, with an opening of $1/10^{th}$ to $1/5^{th}$ of the diameter of the sphere, usually made from metal but possibly from other materials.[26] Its size determines the frequency at which it resonates.

Dunn then builds his theory of what once existed inside the Grand Gallery.

---

[23] Dunn, op. cit., p. 134.

[24] Ibid., p. 164.

[25] Ibid., p. 165.

[26] Ibid.

To extrapolate further we could say that each resonator assembly that was installed in the Grand Gallery was equipped with several Helmholtz-type resonators that were tuned to different harmonic frequencies. In a series of harmonic steps, each resonator in the series responded at a higher frequency than the previous one....To increase the resonators' frequency, the ancient scientists would have made the dimensions smaller, and correspondingly reduced the distance between the two walls adjacent to each resonator. In fact, the walls of the Grand Gallery actually step inward seven times in their height and most probably the resonators' supports reached almost to the ceiling. At their base, the resonators were anchored in the ramp slots.[27]

He then produces the following diagrams of the resonator assemblies arrayed in the Grand Gallery.

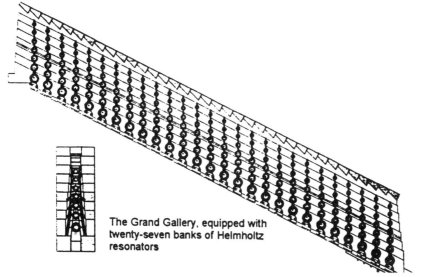

The Grand Gallery, equipped with twenty-seven banks of Helmholtz resonators

Let us pause at this juncture to observe some important points. First, note that Dunn has independently corroborated what Sitchin's texts indicate: that certain components, crucial to the functioning of the Pyramid, once existed inside the Grand Gallery itself. However, note also that there is a contradiction between what Sitchin's texts state and what Dunn hypothesizes once existed in the Grand Gallery. Sitchin's texts refer to the Gallery as having

---

[27] Ibid., p. 166.

been bathed in multi-colored light from several "magic stones" or crystals arrayed in the notches on the side ramps. Dunn, conversely, and on the basis of sound engineering principles, concludes that the primary function of the Grand Gallery was as an acoustic harmonic amplification chamber. In the next chapter, we will present a speculative resolution of this contradiction.

5. The Antechamber: Sound Baffle

With this very sound hypothesis in hand, Dunn next tackles the Antechamber. Drawing on Borchardt's hypothesis that the three slots did indeed once contain slabs that were lowered like a portcullis, he presents a credible theory of why such a machine would have been necessary. His solution is elegant. Whatever was raised or lowered in the slots in the antechamber functioned to block out sound waves coming from the Grand Gallery harmonic amplification chamber that were not of the desired frequency. By raising or lowering these objects, "sound waves with an incorrect frequency have wavelengths that do not coincide with the distance between the baffles and are filtered out."[28] Only the desired frequencies actually reach the King's Chamber.

6. The Air Shafts: Microwave Input and Output

When turning to the King's Chamber, there are three features that are the central focus of Dunn's attention. First, that the chamber itself resonates to the note f# on our own musical scale, a note that is a harmonic overtone of the earth's own Schumann resonance. Second that the airshafts are not for the purpose of emitting air at all. And finally, the Coffer serves the purpose of coupling the input from the "air shafts" with the acoustical harmonic amplification coming from the Grand Gallery.

> In this power plant the vibrations from the earth cause oscillations of the granite within the King's Chamber, and this vibrating mass of

---

[28] Ibid., p. 174.

196

igneous, quartz-bearing rock influences the gaseous medium contained within the chamber. Currently this gaseous medium is air, but when this power plant operated, it was most likely *hydrogen* gas that filled the inner chambers of the Great Pyramid. The Queen's Chamber holds evidence that it was used to produce hydrogen...To maximize the output of the system, the atoms comprising this gaseous medium contained within the chamber should have a unique characteristic – the gas's natural frequency should resonate in harmony with the entire system.[29]

It takes little imagination to understand the significance of using hydrogen as the gaseous medium inside the King's Chamber. Moreover, its presence there would explain the obvious melted look to the Coffer *if at some point an accident or deliberate destruction occurred inside the chamber.*

As to how it all worked, Dunn puts it this way.

Based on the previous evidence, sound must have been focused into the King's Chamber to force oscillations of the granite, creating in effect a vibrating mass of thousands of tons of granite. The frequencies inside this chamber, then, would rise above the low frequency of the Earth – through a scale of harmonic steps – to a level that would excite the hydrogen gas to higher energy levels. The King's Chamber is a technical wonder. It is where the Earth's mechanical energy was converted, or transduced, into usable power. It is a resonant cavity in which sound was focused. Sound roaring through the passageway at the resonant frequency of this chamber – or its harmonic – at sufficient amplitude would drive these granite beams to vibrate in resonance. Sound waves not of the correct frequency would be filtered in the acoustic filter, more commonly known as the Antechamber.[30]

Thus the hydrogen gas would be both acoustically and electrically stressed or pulsed. As the hydrogen atoms absorbed this energy, their electrons would be pumped to a higher state, and undergo quantum jumps until they fell back to their ground state. As they

---

[29] Ibid., p. 179, underlined emphasis in the original, italicized emphasis added.

[30] Ibid., p. 183.

did so, they would release a packet of energy in the microwave region of the electromagnetic spectrum.

7.   The Coffer: The Optical Cavity of a Maser

Dunn notes that one of the peculiar features of the Coffer, itself an object with so many dimensional ratios to the earth, the solar system, and the galaxy, is also an optical cavity with concave surfaces at each end.   Since the electrons of the hydrogen atoms can be stimulated to fall back to their ground state by an input signal of the same frequency, one has here all the makings of a maser: signal input, an optical cavity to cohere the emission of photons as electrons jump out of and back to their ground state, and, in the southern "air shaft" leading from the King's Chamber back to the face of the Pyramid, a horn antenna used to collect microwave beams. Thus the airshafts are not airshafts at all, but waveguides for microwave signal input and output. Dunn thus reasonably assumes, on sound scientific principles, that the Coffer was once correctly positioned exactly between the two shafts.[31] "The (originally smooth) surfaces on the outside of the Great Pyramid are 'dish-shaped' and may have been treated to serve as a collector of radio waves in the microwave region that are constantly bombarding the Earth *from the universe*. Amazingly, this waveguide leading to the chamber has dimensions that closely approximate the wavelength of microwave energy, 1,420,405,751,786 hertz."[32] This is tantamount to saying that the Pyramid's engineers built a structure that was designed to collect the background radiation of the universe, radiation that most physicists currently believe was left over from the "Big Bang" itself, and that plasma cosmology maintains is the result of the electromagnetic vorticular processes at evidence in galactic structures.

---

[31] Ibid., pp. 184-185.
[32] Ibid., p. 186, emphasis added.

8.   The Queen's Camber: A Hydrogen Generator

On what basis does Dunn assume that hydrogen was indeed the gaseous medium inside the King's Chamber? Dunn is unhesitating in his belief that hydrogen was the gas used to power the Pyramid. "Without hydrogen this giant machine would not function."[33] Noting that early explorers to the Queen's Chamber beat a hasty retreat because of its overpoweringly unpleasant odor, Dunn speculates that a chemical reaction, such as between zinc and hydrochloric acid, were used to produce the hydrogen gas. Other chemical processes may have been used incorporating hydrogen sulfide, which would account for the odor.[34]

9.   Meltdown, or Deliberate Destruction?

Having constructed this complex theory, Dunn then proceeds to account for the apparent violent disruption and dislocation evident in the King's Chamber: the slanted cracked walls, the melting of the Coffer, and the blackened limestone face on the surface of the interior of the Grand Gallery. These things he attributes to a "malfunction" that led to the hydrogen, "for some inexplicable reason," exploding in a ball of fire.[35]   Having undergone this accident, he theorizes that the Pyramid's builders then tunneled up to the Grand Gallery in order to make repairs, the occasion of cutting the much-debated "well shaft."

Here again, however, Dunn's theory contradicts the paleographic evidence marshaled by Sitchin. In the version preserved in the ancient texts, the Pyramid was entered for the deliberate purpose of inventorying its contents, and for earmarking some components for destruction and others for removal to be used in other devices elsewhere. In the face of clear evidence that there was some catastrophic destruction that took place on the inside of the Pyramid in the King's Chamber, and in the face of the

---

[33] Ibid., p. 191.
[34] Ibid., pp. 200, 195.
[35] Ibid., p. 209.

paleographic testimony that this destruction was deliberate, I believe it is safe to say that the evidence of that destruction itself is the strongest corroboration of the textual evidence that states its primary function was as a weapon.

10. The Other Pyramids and Tesla

Dunn is alive to more sinister uses to his power plant theory, among them Tesla's use of pulsed harmonic vibrations.

> By applying Tesla's technology in the Great Pyramid, using alternating timed pulses at the apex of the pyramid and in the Subterranean Chamber – a feature, by the way, that all the Egyptian pyramids have – we may be able to set into motion 5,273,834 tons of stone! If we have trouble getting the Great Pyramid going, there are three small pyramids nearby that we can start first to get things moving.[36]

These pyramids, he hypothesizes, may have been employed "to assist the Great Pyramid in achieving resonance."[37]

But these insights, made almost in passing, raise as many questions as they answer, and in doing so, point out the relatively few weaknesses in Dunn's brilliant analysis. Once having mentioned Tesla, Dunn does not really go into any significant detail about how Tesla's work may have been utilized in the Great Pyramid other than to set up vibrations. And if the other pyramidal structures at Giza were designed to help the Great Pyramid "achieve resonance", one must ask, resonance with what? It was, after all, *already* resonant to the earth. Perhaps their function, then, remains to be discovered in another direction entirely. Finally, while Dunn is aware of them, he never entirely satisfactorily delves into the reasons for all the Pyramid's *other* properties, its alignments with various celestial bodies. Suggestively, however, he does speculate at the end of his work

---

[36] Ibid., p. 149.
[37] Ibid., p. 219.

that the Pyramid's builders knew quite about more about the control of gravity than we do.[38]

It remains to be seen whether a different understanding of all these things is plausible. We will examine the Weapon Hypothesis in chapter seven. But first, there is an astonishing surprise in the ancient texts, and in the Pyramid itself, that we must investigate.

---

[38] Ibid., pp. 253-254. Dunn always refers to the Pyramid's builders as Egyptians/

VII.

The Paleography of Paleophysics, Part Two:
Pythagoras, Plato, Planck, and the Pyramid

"To the man who pursues his studies in the proper way, all geometric constructions, all systems of numbers, all duly constituted melodic progressions, the single ordered scheme of all celestial revolutions, should disclose themselves... (by) the revelation of a single bond of natural interconnection."
Plato, <u>Epinomis</u> 991e, 992a

A. *Tetrahedral Musical Harmonics and Quantum Mechanics in the Pythagorean Plato*

Ernest G. McClain has presented perhaps the most thorough and persuasive argument that a sophisticated paleophysics once existed that was passed down in the coded myths of ancient secret societies. His magisterial treatise, *The Pythagorean Plato: Prelude to the Song Itself*, is a well-argued case that the fundamental mathematical and physical laws of equal musical tempering, the basis of our modern western musical system and its twelve equidistant chromatic tones,[1] were encoded in detailed descriptions of allegorical passages of Plato. As McClain notes,

> When Plato died in 347 B.C. his pupils and friends immediately began to argue about these mathematical constructions and about Plato's purpose in using them for models of souls, cities, *and the planetary system.* By the beginning of the Christian era, much of Plato's mathematics had become a riddle....
> Down through history Plato's mathematical allegories defied Platonists either to reconstruct his arithmetic or to find in it the implications he claimed for it.[2]

---

[1] These are the tones one may find on any piano or organ keyboard in any given octave.

[2] Ernest McClain, *The Pythagorean Plato: Prelude to the Song Itself,* p. 1.

McClain's detailed analysis of musical equal tempering in Plato is fully persuasive, but it raises a much more profound set of questions:

- Why would the motions of the planets be coupled to such a system?
- Why would Plato have gone to such great lengths to encode such a system? Why go to such lengths merely to encode "equal tempering" unless that system had something to do with something far more profound than merely a musical system?

The last question suggests that something much more important, a tremendous secret of the ancient unified paleophysics, was at stake. In this section we shall summarize McClain's presentation in an attempt to uncover that secret. Briefly stated, that secret is:

- That the "equal tempering" musical harmonic code encrypted in the Platonic mathematical allegories is only the *first* layer of a much more complex physics found encoded in Plato. McClain has explored only that first layer;
- That harmonic multiples of Planck's constant, the Planck length, and the Planck mass are expressed as *acoustic* information;
- That this information occurs in some cases precisely at the tetrahedral hyperdimensional angles of ~19.5° ±1°; and,
- That these insights allow the broad outlines of a tetrahedral hyperdimensional physics model of systems kinetics to be reconstructed.

Once these components of the ancient paleophysics are recovered, one is then in a position to speculate on the engineering of the components of the Giza Death Star, including the missing components and their possible functions.

McClain observes that the Platonic scholar Robert Brumbaugh

> Noted that the principled of 'aesthetic economy' in Pythagorean use of smallest integers – for examples of general relations in number theory – is itself a purely logical device in an age which (sic.) had not yet developed a general notation for algebraic variables. He noted the

importance for Plato of the circle as (a) *cyclic* metaphor involving 'some sort of reciprocity.'"[3]

That is to say, the use of these numbers actually represents an arithmetic technique of what modern mathematicians and physicists call "harmonic analysis."

> Harmonic analysis is the study of objects (functions, measures, etc.), defined on topological groups. The group structure enters into the study by allowing the consideration of the translates of the object under study, that is, *by placing the object in a translation invariant space.* The study consists of two steps. First: finding the 'elementary components' of the object, that is, objects of the same or similar class, which exhibit the simplest behavior under translation and which 'belong' to the object under study (harmonic or spectral analysis); and second, finding a way in which the object can be construed as a combination of its elementary components (harmonic or spectral synthesis).[4]

One notes that the Platonic "arithmetical analysis of harmonics" is intended to be "translation invariant" because:

- Plato claims for it that harmonics are the basis of planetary motions;
- Because he uses it in connection with the much smaller motions of music; and,
- Because these arithmetic laws also embody motion and action *at the quantum scale.*

The result of a careful analysis of this "arithmetized harmonics" is a system that

> None of us could have anticipated: not only are all of Plato's mathematical allegories capable of a musical analysis – one which (sic.) makes sense out of every step in his arithmetic – but all of his allegories taken together prove to be a unified treatise on the musical scale so that each one throws light on the others.[5]

---

[3] Ibid., p. 2.

[4] Yitzhak Katznelson, *An Introduction to Harmonic Analysis* (Dover, 1976), p. vii.

[5] McClain, op. cit., p. 3.

That the ancient paleophysics should have placed such emphasis on acoustic phenomena and harmonics is not surprising, since they are the first physical laws besides astronomy to have been mathematically modeled.[6] As we shall discover in the next chapter, however, there is a more profound connection between acoustics and gravity.

The problem of equal tempering is basic to this physics and to the engineering of it. Today we divide the musical octave into twelve equal parts with the value of $^{12}\sqrt{2}$. This equal tempering gives the following scale:[7]

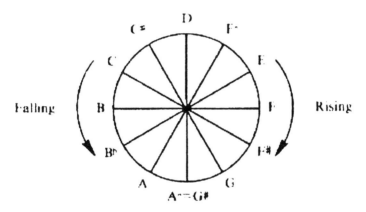

Figure 1:
The Equal Tempered Scale

However, musicians know that an octave of the ratio 1:2 is not divisible by ratios of pure rational numbers, because the powers of even numbers (2,4,8...etc) that define octaves are *never* coincident with powers of 3(9,27,81, etc.) that define intervals of fifths and fourths. Additionally, neither of these overtone series are coincident with the powers of 5 that define the intervals of thirds. Cyclic agreement or unification of these three overtone series *can only be accomplished by deliberate deformation of the intervals on*

---

[6] Ibid.
[7] All diagrams are McClain's.

the basis of a close approximation of $^{12}\sqrt{2}$. That is to say, equal tempering is the first known example in theoretical physics of the "unification of fields," in this case, the fields of information constituted by the three overtone series of octaves, fifths and fourths, and thirds. It is to be noted that such unification is achieved by *engineering*, i.e., by the deliberate *distortion and close approximation* of the "pure" relationships of absolute mathematical and physical theory. Not to approximate these relationships would lead to the "harmonic chaos" of an infinite number of overtones to a fundamental.[8] And that in turn provides a clue as to how the paleoancient Very High Civilization may have achieved a unified physics.

The basis of this encoded Platonic equal tempering is the harmonic ratio that Pythagoras allegedly brought to Greece from Babylon. The implications of this allegation should be obvious, for it tends to corroborate the supposition of a paleoancient Very High Civilization of which Sumeria and Greece are the considerably declined legacies. The musical ratios that Pythagoras brought is the ratio

$$6:8::9:12.$$

Taking this ratio to define the octave space, it has two means, the arithmetic mean $M_a = 1\frac{1}{2}$ and the harmonic mean $M_h = 1\frac{1}{3}$:

$$
\begin{array}{cc}
M_h & M_a \\
6:8 \; :: \; 9 : 12 \\
\\
3:4 \quad 3:4 \\
\\
2 \;\; : \;\; 3 \\
\\
2 \;\; : \;\; 3
\end{array}
$$

---

[8] McClain, op. cit., p. 4.

These proportions apply both to rising and to falling pitch sequences:

	6:	8 ::	9 : 12
Rising	D	G	A D
Falling	D	A	G D

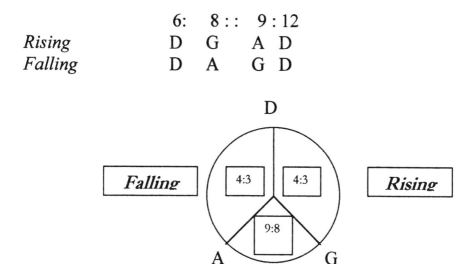

Plato states:

> (The legislator) must assume it as a general rule that numerical division in all its variety can be usefully applied to every field of conduct. It may be limited to the complexities of arithmetic itself, or extended to the subtleties of plane and solid geometry; it is also relevant to sound and motion, straight up or down or revolution in a circle.[9]

Note that Plato has said an astonishing thing: *every field of human conduct or investigation may be mathematically modeled.* Thus physics may be arithmetically and mathematically modeled, a model determined by arithmetic, harmonic, and geometric means.

In this regard, we turn to a consideration of one of the most important components of McClain's analysis, the mystifying Platonic "sovereign number" of 60^4, or 12,960,000. McClain notes

---

[9] Plato, *Laws* 747a, cited in McClain, p. 10.

that the function of this number in Platonic Harmonics is to be a "tonal index", that is, "an arbitrary terminus for the potentially endless generation of tone-numbers, a limitation which (sic.) provides integer expressions for some set of ratios."[10]

However, there is nothing arbitrary about this number whatsoever, *because this number is divisible by harmonics of Planck's constant, by harmonics of the Planck Length, and by harmonics of the Planck Mass simultaneously and to within one decimal place.* Taking as the theoretical value of these numbers to within three decimal places, and converting metric to English measures since the latter so closely approximate the units of measure in the Great Pyramid, one obtains the following harmonic numbers of the Planck units:

Planck Unit	Theoretical Value	Harmonic Number
ħ (Planck's constant)	6.626×10^{-34} joules	6626[11]
Ł (Planck length)	6.362×10^{-8} inches	6362
M_P(Planck mass)	4.799×10^{-8}	4799

Dividing Plato's "sovereign number" 12,960,000 by the harmonic values of the Planck units, one obtains a rather astonishing and breathtaking result:

Platonic Number	÷ 6626	÷6362	÷4799
12,960,000	1955.931	2037.095	2700.562
	(1956)	(2037)	(2700 or 2701).

I will call these numbers alternatively the "sovereign harmonics" or "Planck harmonics" for they may in turn be divided or multiplied by the four numbers of the ratio 6:8::9:12.

---

[10] McClain, op. cit., p. 17.

[11] Planck's constant has a theoretical value of 6.626076×10^{-34} joules.

Planck Harmonic	6	8	9	12
OF PLANCK'S CONSTANT (1956)	*11,736* 326	*15,648* 244.5	*17,604* 217.33	*23,472* 163
OF PLANCK LENGTH (2037)	*12,222* 339.5	*16,296* 254.6	*18,333* 226.33	*24,444* 169.75
OF PLANCK MASS (2700)	*16,200* 450	*21,600* 337.5	*24,300* 300	*32,400* 225

These close approximations are significant because they tend not only to confirm that Plato was indeed an initiate into the "Egyptian mysteries", but also because those mysteries apparently encompassed a physics that was not only harmonic but sophisticated enough to know the fundamentals of nuclear and quantum mechanics. This implies that someone at some time in the remote past intended to preserve that knowledge and engineering for a future time. Given that the Great Pyramid was a weapon of mass destruction, it also implies that this "someone" also intended not only to preserve the knowledge but the purpose to which that knowledge was employed: weaponry.

Returning now to McClain's exposition of Plato, the "perfect number" for Plato is 6, since it is the sum of its proper divisors 1,2,3. Thus the ratios of the first six integers 1:2:3:4:5:6 define the tones of the Greek Dorian mode and "its reciprocal, our modern major scale."[12]

---

[12] McClain, op. cit., p. 20.

Greek Dorian	D	c	b^b	A	G	f	e^b	D	*(Falling)*
Reciprocal	D	e	$f^{\#}$	G	A	b	$c^{\#}$	D	*(Rising)*
Ratios	1			:				2	

$$2 \quad : \quad 3 \quad : \quad 4$$
$$4 \; : \; 5 \; : \; 6$$
$$4 \; : \; 5$$
$$5 \; : \; 6$$
$$4 \; : \; 5 \; : \; 6$$

Plato then makes an astounding statement in the *Laws* that clearly indicate that he is indeed talking about how the visible cosmos comes into existence via harmonics that arise from a smaller quantum and *sub*-quantum substrate:

> But the condition under which coming-to-be universally takes place-what is it?
> Manifestly 'tis effected whenever its starting point has received increment and so come to its second stage, and from this to the next, and so by three steps acquired perceptibility to percipients.[13]

The three steps of this sub-quantum systems kinetics will be explored more fully in the next section. Most Platonic scholars agree that what Plato had in mind in this passage was the Pythagorean tectractys:

Point   **(1)**

Line   A   B

Plane A^2 AB B^2

Solid A^3 A^2B AB^2 B^3

---

[13] Plato, *Laws* 894.

Given the encoding of sophisticated quantum mechanics found thus far, one may speculate that the Pythagorean tectractys may be a model for our own gauges of scale:

Point				(1)				Aether?
Line			A		B			sub-quantum
Plane		A^2		AB		B^2		particle?
Solid	A^3		A^2B		AB^2		B^3	atom?

The allusion to models of hyperdimensional physics such as supersting theory is further corroborated in that Plato considers the number 10, the number of nodal points in the tectractys, as his limit to "form numbers" and that he also considers it as a "time factor."[14] 10 is the number of dimensions underlying reality in one rotation of string theory, with 4 the number of dimensions in the "real" world with a further 6 dimensions "curled up inside" of it.[15]

Point	•	Aether	IMPLICATE
Line	• •	Sub-quantum	ORDER
	• • •	Particle	
Solid	• • • •	Atom	EXPLICATE ORDER

The difference of the harmonic series of these two systems with respect to each other then leads to the problem of the "Pythagorean Comma" in the Platonic exposition of equal tempering. Taking the 9:8 ration of the 6:8::9:12 musical ratio, and rotating the two systems in contrary directions, one obtains

---

[14] McClain, op. cit., pp. 42-43.
[15] Michio Kaku,

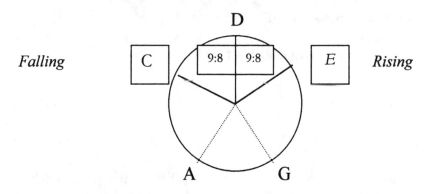

If "D" is taken as the harmonic "center of symmetry" for the two systems, or as its "base condition" or "base time", calculating by powers of 9/8 yields a discrepancy at $G^{\#}$ and A^{b} whose values should be the same:[16]

$8^{6} = 262,144 = A^{b} = 512^{2} = 2^{18}$; x 2 = 524288
x 9/8 = 294,912 = B^{b}
x 9/8 = 331,776 = C
x 9/8 = 373,248 = D (the harmonic center of symmetry)
x 9/8 = 419,994 = E
x 9/8 = 472,392 = $F^{\#}$
x 9/8 = 531,441 = $G^{\#}$ = 729^{2} = 9^{6}.

Thus, the notes $G^{\#}$ and A^{b}, which on our keyboards are the same, are *not* the same in the natural harmonic series ascending and descending from the note D:

---

[16] McClain, op. cit., pp. 36-38.

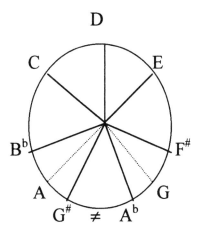

This ratio of 524288 : 531441 is the "Pythagorean Comma," a ratio of approximately 73:74.[17] If one subtracts the difference between the two numbers of the ratio one obtains 7153 and dividing by the Planck mass gives 14.905188. An acoustic relationship between gravity and harmonics would seem to be suggested.

This is further confirmed if, as McClain suggests, the Pythagorean comma is extended around the circle. The comma will be reproduced at three places, $g^{\#}:a^{b}$, C:c, and E:e. McClain then produces a diagram of these relationships (cf. next page).

---

[17] McClain, op. cit., p. 37.

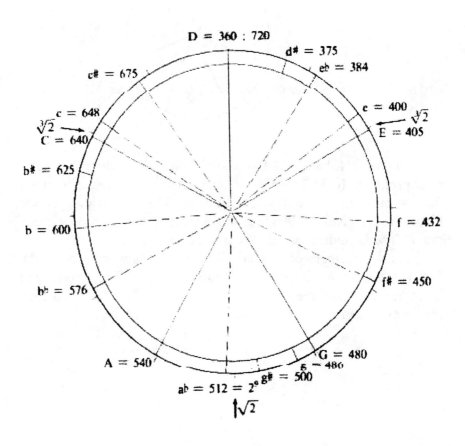

Once again, this diagram contains some astonishing approximations of whole number multiples of the Planck units. Taking the values of b^b, c, D, e, $f^\#$, and a^b, multiplying them by 100, and dividing by the Planck units reveals some unusual multiples:

NOTE	MULTIPLE	M_P	Ł	ħ
D	36,000	7.5015628	--	--
e	40,000	--	6.287331	--
$f^\#$	45,000	--	7.0732474	--
a^b	51,200	--	8.0477837	--
b^b	57,600	12.0025	9.0537566	--
c	64,800	--	10.185476	9.7796559

I have indicated only the closest approximations.

Another series of relationships may be found in another diagram produced by McClain. Significantly, at *every* important angle for tetrahedral hyperdimensional physics some harmonic of a Planck unit may be found.[18]

NOTE	HARMONIC	DIVIDED BY:		
		M_P	Ł	ħ
D^o	50,400	1.502	7.922	--
$e^{\wedge 1}$	28,800	6.001	--	--
g^{bv2}	31,360	--	4.929	--
$a^{\#\wedge 2}$	40,500	--	--	6.112
c^{v1}	44,100	9.189	6.931	--

Plato's mathematical trigonometry generates a similar, and quite extensive table of numbers. The rules for generating triangles in Pythagoreanism are that any two numbers {p,q}, whith p>q, generates triangles as follows. If p=2 and q=1, we obtain the

---

[18] McClain, op. cit., p. 108.

octave ratio of 2:1, which in turn generates the famous Pythagorean triangle with side ratios of 3:4:5.

Opposite Side
$L=2pq$
$L=2\times2\times1=4$

Adjacent Side
$M=p^2-q^2$
$M=2^2-1^2=3$

Hypoteneuse
$N=p^2+q^2$
$N=2^2+1^2=5$

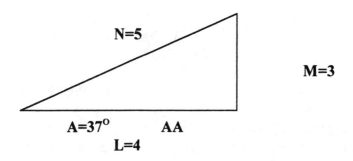

N=5

M=3

A=37°   AA

L=4

It is worth noting that this encoding of the fundamentals of quantum mechanics may not simply be limited to the legacy civilizations – Egypt, Sumer, and Greece – stemming from the remnants of the Very High Civilization that built the Great Pyramid. In chapter three, *The Paleography of Paleophysics*, this rather provocative passage was mentioned:

> The gravity of the centre (sic.) of the earth, the gravity of global earth, the solar flood, the air force, the force emanating from the planets and stars, the sun's and moon's gravitational forces, and the gravitational force of the universe, all together enter the layers of the earth in the proportion of 3, 8, 11, 5, 2, 6, 4, 9 and, aided by the heat and moisture therein, cause the origin of metals, of various varieties, grades, and qualities.[19]

---

[19] David Hatcher Childress, *Vimana Aircraft of Ancient India and Atlantis* (Adventures Unlimited Press: Kempton, Illinois, 1999), p. 16.

Multiplying these numbers together gives 570,240, and dividing by our assigned theoretical harmonic value of 6626 gives 86.060971. It is reasonable to speculate that other non-western esoteric traditions might preserve similar astonishing numerical codes.

More astounding by far, however, are not the approximations to the Planck units in Plato, but in various dimensions of the Great Pyramid itself.

B. Harmonic Approximations of Planck's Constant and the Planck Length in the Giza Death Star

In order to accomplish the unification of gravity, acoustics, and electromagnetism, a sophisticated knowledge of quantum mechanics and the essential measures of the latest theories of quantum gravity, namely, Planck's constant of action and the Planck length should be found in redundant harmonics of those values, *expressed, as the Pyramid expresses so many other of its profound physical and mathematical relationships, as analog approximations present in various dimensions of the components of its structure.* Close and redundant approximations of these quantities or measures will corroborate the view that the Pyramid was the "mechanical" observer of the very effects it weaponized and that one of the purposes of the machine its builders intended for it was the observation and manipulation of quantum and gravitational effects.

1. "Aesthetic Adumbrations" of Planck's Constant in the Great Pyramid

The Great Pyramid's peculiar property of embedding known physical constants and geometrical relationships is bewildering. As physical mechanics advances, more perplexing discoveries are made about the structure, which seems to embed these discoveries in terms of the mathematical measures of certain of its dimensions expressed in units of measure that are themselves embedded in the structure.

If the Great Pyramid is a harmonic oscillator representing a sophisticated weaponization of a unified physics, as I have argued in chapter 6, then it must embody some approximation of one of the most important of these constants, Planck's constant of the minimum quantum of action. This is the necessary mathematical constant upon which the development of quantum physics rests.

The Pyramid's analogue approximations of π are well known. In fact, most of the physical constants or geometrical properties both of solar and of local terrestrial space find some expression in the Pyramid in terms of redundant *approximations*. It stands to reason that one may expect similar approximations of this crucial physical constant if in fact the Pyramid was engineered on the basis of a sophisticated unified physics. Planck's constant has a theoretical value of 6.626076×10^{-34} joules, which for our purposes here we round off to 6.626×10^{-34} joules. Expressed in terms of a harmonic number thirty-seven orders of magnitude greater, we may concentrate on the value 6626. Similarly we may again take approximations of the Planck Length as 6362 and the Planck Mass as 4799.

I argue in the next chapter that the Grand Gallery may have been a "gravito-acoustic" resonance and amplification chamber filled with now-missing artificial and possibly liquid crystals that were both electro-magnetically, gravitationally, and acoustically resonant. One finds a weird "aesthetic" adumbration of Planck's constant in the measure of the length of the Great Step, from the end of the Grand Gallery, to the passage leading to the Antechamber. Expressed in Pyramid inches this dimension is 61.6266.[20] Perhaps this "aesthetic" resemblance at such a crucial place in the structure of the Pyramid and the hypothesized functions ascribed to them in my hypothesis is indeed significant.

Unfortunately, mere adumbrations and resemblances are not sufficient. Rigor does not permit us to speculate on why this resemblance exists, but only to mention it. A purposeful

---

[20]E. Raymond Capt, M.A., A.I.A., F.S.A, SCOT., *Study in Pyramidology* (Thousand Oaks, California: Artisan Sales, 1986), p. 88.

embedding of Planck's constant would require *redundant* examples, and that these examples occur in places that would tend to confirm our hypothesis of a unified physical mechanics. Are there indications that this occurs, and that its builders were in fact in possession of such a unified physics *that was capable of being engineered*?

Taking the theoretical harmonic value for the constant, or 6626 = ℏ, one discovers that $ℏ^2 \cong 43903876$ which is also an approximate harmonic expressed in miles of the neutral or transition point between the earth and the moon, where an object is under equal gravitational attraction from the two bodies. The local planetary lunar-terra system would thus seem to be expressed in terms of a harmonic of the quantum of action represented by $ℏ^2$. This would appear to corroborate our initial hypothesis of the extraordinary degree to which the ancient paleophysics of harmonic systems entanglement was unified and scale invariant.

And this indicates the "adumbration" may be something more. 10ℏ would equal 662600, and subtracting the distance measure of the Great Step to the Antechamber of 616266 gives 46334, which is approximately the value of 7ℏ (46383). Why is this significant? First, because it gives another redundancy. And second, because the measures are all approximations of the neutral point between the moon's and the earth's gravity. *But one can only have this mean point if the gravitational attraction of the moon is significantly larger than the $1/6^{th}$ gravitational attraction usually ascribed to the moon.* A discrepancy exists between published figures for the moon's gravity and the mean point between the earth and the moon's gravitational fields.[21]

Based on calculations for the mean point based on Newtonian mechanics and the idea of the moon having $1/6^{th}$ the gravity of the

---

[21] Cf. William L. Brian II, *Moongate: Suppressed Findings of the U.S. Space Program: The NASA-Military Cover-up* (Portland, Oregon: Future Science Research Publishing Co., 1982), pp. 29-60. Brian points out the great discrepancies between pre-Apollo and post-Apollo calculations for the neutral point. Hoagland likewise maintains *that* we went to the moon, but that some sort of cover-up occurred having to deal with the physics that got us there and back.

earth, the mean point would have been much closer to the moon. This error may have led to early disasters in sending probes into lunar orbit, since miscalculation of this mean point meant that orbital insertions occurred at the wrong point and velocities. A higher gravitational acceleration on the moon's surface would therefore also imply that far more rocket fuel would have been needed on the LEM when blasting off from the moon than it actually carried. Some other additional means of propulsion must therefore have been required to return the LEM from the moon's surface to the orbiting Command Module. And all of this means two things. First, that there are discrepancies in the moon landing and published facts that suggest a cover-up is taking place.[22] Second, and more importantly, the Pyramid's builders not only knew a great deal about the local gravity of the luna-terra planetary system but were expressing it in terms of two different harmonics of Planck's constant, set to a theoretical value of 6626. In short, they were expressing a planetary system configuration as a harmonic of a quantum state.

An even more interesting embedding of Planck's constant of action is discovered in the measure mentioned by "prophecy-in-stone" exponent Adam Rutherford, who noted that the "Royal Cubit" unit of measure was equivalent to $4y/10^3 \sqrt{\pi}$ sacred cubits, or 0.8242637 Sacred cubits. This is an extremely interesting result, embedding some harmonic of Planck's constant *via the series of numerical and geometrical relationships that Richard Hoagland has discovered in the Cydonia complex on Mars.* If one takes $\sqrt{5}$, with a theoretical value of 2.235, dividing it by the constant ε with a theoretical value of 2.716, one obtains 0.823. $\hbar/8$ yields a similar result, to two decimals, of 0.828.

While these results do not constitute anything like an established case, they are suggestive in that they *tend to confirm* the type of physics being posited as the basis of the Pyramid's

---

[22] I wish to make clear that I do *not* dispute that the moon landings took place, only that there are aspects of the landings not adequately explained in the publicly released information.

engineering and the purpose that motivated the engineering. The "aesthetic adumbration" and the two analog approximations of 6626 suggest that further redundancies might be discovered. Since approximations occur in the expression of a planetary relationship expressed as a harmonic of quantum action, one may also expect that if further redundancies are discovered they will be found in similar planetary geometries that are in turn duplicated in the Pyramid.

2. Unexpected Redundant Approximations and Harmonics of Planck's Constant In Significant Functional Structures of the Great Pyramid.

In the next chapter I hypothesize that the basic unified physics that was weaponized in the Great Pyramid utilized some engineered version of Bohm's "pilot wave"[23] acting as a carrier wave for harmonically coupled and cohered acoustic and electromagnetic energy. Carried by such a superluminal pilot wave, energy travels in non-linear, time-reversed fashion directly into the nuclei of targets, pulsing energy past the threshold of stability and initiating nuclear reactions. Direction of such pulsed "pilot waves" would seem to occur by dint of some form of harmonic interferometry, which, I suggest in the next chapter, was the possible function of some of the other structures at Giza.

Given this basic hypothesis, and reliant upon the careful analysis of Dunn and Sitchin, the four principle internal chambers within the Great Pyramid – the Queen's Chamber, the Grand Gallery, the Antechamber, and the King's Chamber – and their missing components *should* exhibit redundant harmonic approximations or multiples of these Planck units.

A word is necessary as to why this is so. The Great Pyramid itself is not only a harmonic oscillator (Dunn's hypothesis), and

---

[23] One may, perhaps, speculate that this notion to maintain that Bohm's "pilot wave," Tesla's "standing wave" and Bearden's "scalar waves," the probability wave in the phase space of quantum mechanics belong to the same family of wave forms.

not only a radiation collector and reflector (its parabolic polished faces), but a huge *crystal*. Careful study of the dimensions of its internal chambers reveals that it is also a harmonic oscillator composed of several other harmonic oscillators, each focusing and amplifying their energies into the King's Chamber. In short, the Pyramid is constructed as a series of nested feedback loops.

The functions of these four chambers that then suggest themselves – and anticipating somewhat the results of the next chapter are these:

(1) The Queen's Chamber was understood to have housed some mechanism for the induction of chemical reactions to create hydrogen gas, and possibly to initiate some endothermic plasma state in that gas pace Dunn.

(2) The Grand Gallery is a gravito-acoustic amplification chamber, filled (perhaps) with artificial (possibly liquid) crystals resonant acoustically and gravitationally with the earth and other celestial systems in a manner similar to Dunn's Helmholtz resonator arrays. On this analysis, we accept Dunn's conclusion that the chamber was filled with hydrogen, and in our view, with an endothermic hydrogen plasma that provided most of the nuclear and electromagnetic energy being coupled and amplified in the chamber.

(3) The Antechamber – again following Dunn – is understood to be a baffle chamber filtering non-resonant wavelengths, i.e., wavelengths not resonant with the target, from entering the King's Chamber.

(4) The King's Chamber is understood to be a complicated phase conjugate mirror and howitzer, relying upon the principle of harmonic coupling of all known forms of energy and cohering that coupled energy to the superluminal pilot wave.

As will be evident, these structures are precisely where one would most expect redundant approximations or Planck's constant or some harmonic thereof. Indeed, when one *does* turn to certain measures of dimensions within these chambers, there are significant approximations and redundancies.

(1) Redundancies in the Queen's Chamber

There are four significant redundant harmonic approximations of Planck's constant in the Queen's Chamber. Expressed in Pyramid Inches (PI) these four measures are:

(1) Width of niche at bottom (north to south)	61.81978 PI
(2) Width of niche at top (north to south)	20.60659 PI
(3) Depth of niche (east to west)	41.21319 PI
(4) East-west Distance from Queen's Chamber Passage axis to North-South niche axis	41.21319 PI

Taking harmonic values of these numbers and dividing by the theoretical value of 6626 for h, we obtain the following:

Harmonic Number of Measure Harmonic of h

(1) 6181978	932.98792 h
(2) 2060659	310.99592 h
(3) 4121319	621.992 h
(4) 4121319	621.992 h

(2) Redundancies in the Grand Gallery

There are six such redundancies in the Grand Gallery:

(1) Width between ramps	41.2139 PI
(2) Width of roof	41.2139 PI
(3) Width over top of ramps	82.42637 PI
(4) Length of roof (approx)	1836.000 PI
(5) Distance across top of Great Step(north to south)	61.62660 PI
(6) Distance across top of Great Step(east to west)	82.42637 PI

These values compute to the following harmonics of h:

Harmonic Number	*Harmonic of ħ*
(1) 412139	621.992 h
(2) 412139	621.992 h
(3) 8242637	1243.9838 h
(4) 18360000	2770.9025 h
(5) 6162660	930.07244 h
(6) 8242637	1243.9838 h

Redundancies in the Antechamber

There are ten redundancies in the Antechamber, three in the first low passage section, five in the Antechamber proper, and two in the second low passage section.

Passage – First Low Section

Dimension	Measure
(1) Height	41.21319 PI
(2) Width	41.21319 PI
(3) Length	52.02874 PI

Harmonic Number	Harmonic of h
(1) 4121319	621.992 h
(2) 4121319	621.992 h
(3) 5202874	785.22094 h

(a) Antechamber Proper

Dimension	Measure
(1) Floor Width	41.21319 PI
(2) Length of granite part of floor	103.03296 PI
(3) Height of east wainscot	103.03296 PI
(4) Height of Granite Leaf above floor	41.21319 PI
(5) Distance from face of boss to north wall	20.60659 PI

Harmonic Number	Harmonic of h
(1) 4121319	621.992 h
(2) 10303296	1554.9797 h
(3) 10303296	1554.9797 h
(4) 4121319	621.992 h
(5) 2060659	310.99592 h

(b) Passage – Second Low Section

Dimension	Measure
(1) Height	41.21319 PI
(2) Width	41.21319 PI

225

Harmonic Number	*Harmonic of h*
(1) 4121319	621.992 h
(2) 4121319	621.992 h

(3) Redundancies in the King's Chamber

There are seven redundancies in the King's Chamber, with two found in the Coffer.

Dimension	*Measure*
(1) Length east to west	
(2 x 365.24235/√π)	412.13186 PI
(2) Width north to south (365.24235/√π)	206.06593 PI
(3) Height (√5 x 365.24235/√π)	230.38871 PI
(4) Floor Diagonal (√5 x 365.24235/√π)	460.77743 PI
(4) Diagonal of east and west walls	
(3 x 365.24235/2√π)	309.09889 PI
(6) Width of Coffer	38.69843 PI
(7) Height of Coffer	41.21319 PI
(sum of Coffer's height, width and	
length = 1/5th King's Chamber's height, width and length	

Harmonic Number	*Harmonic of h*
(1) 41213186	6219.9194 h
(2) 20606593	3109.9597 h
(3) 23038871	3477.0405 h
(4) 46077743	6954.0813 h
(5) 30909889	4664.9394 h
(6) 3869843	584.03908 h
(7) 4121319	621.992 h

Measures (1) – (5) of the King's Chamber are the most significant. In chapter three we saw that two significant features of the unified paleophysics were that time was its primary differential, and that one basic engineering principle embodied in it was the rendering of larger-than-quantum sized targets in terms of some quantum state. Measures (1)-(5) appear to offer some "aesthetic" correlation of these speculations, since the temporal geometry of the earth – its terrestrial year – are expressed in terms of some multiple or harmonic approximation of ħ!

3. Approximations Necessary to Engineer and Weaponize the Physics

Why *only* approximations, however? If the paleophysics was sophisticated enough to be genuinely unified and *engineerable* then one answer – albeit a highly speculative one – suggests itself. The approximations, *expressed in terms of the units of measure unique to the Pyramid itself*, i.e., the "Pyramid inch" and so on, may have been necessary to achieve an *engineerable* unification of physics utilizing standard dimensional analysis. If the Observer - in this case the Pyramid itself - imposed upon the systems it is observing or "harmonically entangling" units of measure arbitrarily selected with no relationship to those systems, then it would not have been possible to "observe," that is, to *couple* them harmonically. The technique is well-known to engineers. While relativistic and quantum mechanics are certainly necessary tools to engineer certain things, most engineers utilize the far simpler mathematics of Newtonian mechanics for most applications and make close approximations in order to achieve "engineerability."

In any case, these redundant approximations, occurring as they do at structurally significant places in terms of the weapon hypothesis, have revolutionary importance. They suggest that the engineers of the Pyramid expressed the makrocosmic in terms of quantum states, and vice versa, that they expressed quantum states in terms of the macrocosmic.

C. Redundant Harmonic Approximations of the Planck Length in The Great Pyramid

Similarly, investigation reveals harmonic approximations or resonances of the Planck Length in the portions of the structure that would necessarily have to do with the structuring of the gravitational and quantum states of the systems being coupled and targeted. Taking the Planck Length of $1.61599 \times 10^{-35th}$ m, and converting to the American inch (a very close approximation to the Pyramid inch), one gets $63.621526 \times 10^{-35th}$ inches. Raising this 33 degrees of magnitude and giving a theoretical harmonic value of 6362 Ł(where Ł as before designates the harmonic Planck Length), one obtains ħ/PL = 1.0415963 and PL/ħ = .9601569.

Remember the Pythagorean Comma that gave a ratio of 73:74 or 74:73? Dividing the numbers of the ratio of the Pythagorean Comma gives astonishingly similar results: 73/74 = .9864864, and 74/73 = 1.0136986.

Again, dividing 6362 into the pure harmonic values of dimensional measures of certain structures where a resonance of the Planck constant is found, the following table is obtained:

Planck Length Harmonics in the Great Pyramid:

Dimension	Measure in Pi	Multiple of Planck Length
Descending Passage		
Vertical Distance of Floor Beginning to Roof Beginning	39.995 (39995)	6.2865451
Entrance to Foot of Scored Lines	481.7457 (4817457)	757.22367
Scored Lines to Intersection Of Floor Lines of Ascending Passage	628.5079 (6285079)	987.9093

PLANCK LENGTH HARMONICS IN THE GREAT PYRAMID:

DIMENSION	MEASURE IN PI	MULTIPLE OF PLANCK LENGTH

Queen's Chamber

Width of Niche at Top (north To south)	20.60659 (2060659)	323.90113
Depth of Niche (east to west)	41.21319 (412139)	647.80242
East-West Distance from Passage Axis to NS Niche Axis	41.21319 (4121319)	64.780242

Grand Gallery:

Width Between Ramps	41.2139 (412139)	64.781358
Width of Roof	41.2139 (412139)	64.781358
Lengtth of Roof	1836.0000 (18360000)	2885.8849
Distance, east-west, across Top of Great Step	82.42637 (8242637)	129.56046

Antechamber:

Distance from Face of Boss To North Wall of Chamber	20.60659 (2060659)	323.90113

King's Chamber

Length East to West $(2 \times 365.24235/\sqrt{\pi})$	412.13186 (41213186)	6478.0235
Width North to South $(365.24235/\sqrt{\pi})$	206.06593 (20606593)	3239.0117

These redundant harmonics both of the Planck constant and Planck length are significant in that both measures are necessary in any quantum theory of gravity, and indicate that gravitational and acoustic, quantum mechanical (nuclear), as well as electromagnetic and optical effects are being engineered in the Pyramid to an extraordinary degree.

D. Time Differentials

In Chapter Three *The Paleography of Paleophysics* I stated that time was the primary differential of the ancient paleophysics, given its preoccupation with harmonics at all scales of size, from the motions of planets to sub-atomic particles. Because of this, I posited a "systems" approach to the Pyramid, since three different systems were coupled and oscillated in it. I call these systems the Base Planetary, Base Stellar or Solar, and Base Celestial or Galactic Systems, designating the earth, the solar system, and the Milky Way galaxy respectively. The motions of these systems, and of the gravitational, electromagnetic, and acoustic forces of planetary and atomic sized bodies may be expressed as "time differentials" in the relative rates of change between each of these systems, since the only way of measuring time is precisely by the relative motions of bodies, whether planetary or atomic.

This leads to a conclusion about one putative principle of paleophysics: *the forces and energies of physics may be modeled or expressed by equations where time is the primary differential, and this in turn may lead to the formulation of a rigorously unified field theory on that basis.*

If these approximations and harmonic multiples of the Planck units in the Pyramid *are* significant – and I believe sufficient redundancy exists to suggest that they are – then what are they telling us?

First, I believe that the ancient paleophysics of the Very High Civilization that built the Giza Death Star was a "praxio-unified" physics as distinct from a *theoretically* unified physics that has no practical engineerability. Thus, the paleophysics, as a "praxio-unified" physics, is distinguished from some aspects of contemporary models such as M (membrane) Theory or Superstring Theory. The "paxio-unified" physics being posited for the paleoancient Very High Civilization indicates the condition of the willingness to settle for analog approximations of the pure mathematical relationships of the theory in order to accomplish a genuinely engineerable application of that physics. Put more bluntly, the paleophysics of the Very High Civilization was unified precisely because it acknowledged the harmonic nature of any attempt at such unification, and because that harmonic unification could only be accomplished through the primacy of engineering, and not mathematical modeling.

And of course, secondly, it indicates that the Great Pyramid – "the most surveyed building in the world" – may not yet have unlocked all its secrets, but only has just begun to do so.

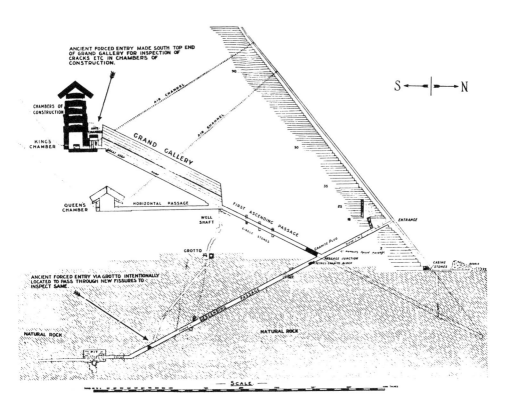

Passage System of Great Pyramid, Looking West.

The Grand Gallery of the Great Pyramid, Looking North (Down). Note first overlap on left exhibits signs of calcination.

PART TWO:
THE INNER CHAMBERS:
THE PULSING HEART OF THE WEAPON

VIII.

The Weapon Hypothesis:
The Weaponized Unified Physics of Harmonic Entanglement of Non-Local Systems

"He can bring down the stars with his arrows, or break apart the continents and let in the sea, or raise earth once drowned and create all Creation anew."
"The Golden Deer", The <u>Ramayana</u>[1]

"And a lightening flash snapped across the sky though there were no clouds ..."
" Ravana and Time," The <u>Ramayana</u>[2]

"There is, of course, an infinite number of possible representations of the path integral. However, as in the point particle case, we can always choose the simplest one, the harmonic oscillator basis.... Notice that each oscillator mode is basically uncoupled from the other oscillator modes."
Michio Kaku, <u>Introduction to Superstrings and M-Theory</u>[3]

A. Marshalling the Evidence: A Review of the Data

A comprehensive review of the evidence is now in order before proceeding to speculate on the type of weapon the Great Pyramid was and on how it might have worked. To aid in the review the evidence is divided into various headings:

(1) Direct Paleographical Evidence of a Weapon Function (Sitchin's evidence),

(2) Corroborative Paleographical Evidence of a Military Aspect of Giza (Hancock's Pyramid Texts),

(3) Indirect Paleographic Evidence of The Use of Weapons of Mass Destruction in ancient Times (The Hindu Epics),

(4) A Comparison of Paleophysics and Modern Physics,

(5) Encoded Harmonics of the Planck units in ancient texts,

---

[1] *The Ramayana,* trans. William Buck (Berkeley: The University of California Press, 1976), p. 163.

[2] Ibid., p. 340

[3] Michio Kaku, *Introduction to Superstrings and M-Theory,* Second Edition (New York: Springer-Verlag, Inc.: 1999), p. 61.

(6) Encoded Harmonics of the Planck units in the Great Pyramid,

(7) Phenomena and Principles of Weaponization Thus Far Suggested, and

(8) Missing Components.

(1) Direct Paleographical Evidence of a Weapon Function (Sitchin's evidence)

Sitchin's texts, cited extensively in chapter two, indicate the following:

- That the Great Pyramid was a weapon of mass destruction.
- Its destructive power exceeded that of nuclear weapons, because his texts also indicate that deliberate care was taken to render it permanently non-operational and because wars were subsequently fought using nuclear weapons. They were not considered to be as terrifying as the Pyramid.
- That the Great Pyramid, when fully operational, gave off strong radiations, requiring the use of protective clothing when entering it.

(2) Corroborative Paleographical Evidence of a Military Aspect of Giza (Hancock's Pyramid Texts)

The *Pyramid Texts* cited by Hancock in chapter two also corroborate a military use and some aspects of the physics principles employed in the weapon:

- The *Pyramid Texts* refer to the compound as the "royal fortress."
- The *Pyramid Texts* also indicate that the "as above, so below" principle was crucial to its operation.
- Once that principle is employed, it gives a "magical protection", indicating a military purpose of at least a defensive nature.

233

(3) Indirect Paleographic Evidence of The Use of Weapons of Mass Destruction in Ancient Times (The Hindu Epics)

The Hindu Epics corroborate, from a different cultural tradition, the very ancient existence and use of weapons of mass destruction:

- Some texts contain descriptive passages of explosions that strongly resemble descriptions of nuclear or thermonuclear detonations.
- Both the Hindu Epics and the vimana texts also contain descriptions of weapons systems that are modern in nature, for example, of aircraft, artillery, missiles, and so on.
- The Hindu Epics also allude to the existence of another kind of weaponry than nuclear bombs of an apparently electromagnetic nature.
- The Vimana texts indicate the existence of sophisticated physical knowledge of the fluid mechanics of the electrodynamics of the layers of the earth's atmosphere.

(4) A Comparison of Paleophysics and Modern Physics

The *Hermetica* shows unmistakable evidence of the existence of a very sophisticated physics:

- The universe is viewed as being a "living" organism, a system of interlocked, mutually reactive cellular structure, a view shared with modern plasma or "hierarchical" cosmology.
- Accordingly, "Soul" permeates space, which is thus viewed as "information in the field", and as such, the soul is capable of instantaneous transference of information. Thus, reality is "non-local", in accordance with Bell's non-locality theorem, and space is not a mere "void" but full of "soul", or information.
- This soul- or information-filled aether requires the presence of an intelligent observer, a view in accordance with the

Anthropic principle of modern theoretical physics, and with some schools of quantum mechanics.

- "Music", that is to say, the harmonic motions of bodies, are the means of entangling non-local systems, since all bodies arise and possess distinction from the aether by dint of the variety of their motions.

- As such, there are harmonic laws of vibration or frequency to which all bodies are subject. This basic principle is true regardless of the size or scale of the object. Such laws – while currently unknown – appear to be corroborated by advances in the field of plasma cosmology that suggest there are electromagnetic laws that are scale invariant, i.e., that apply from the laboratory experiment with plasmas to galaxies themselves. This is to say that all levels of physical reality from the quantum to the galactic scale operate according to the principles of the same geometry.

- Since motion, frequency, vibration and harmonics are the basis of this physics, every object has a "time lock" or "base time" vis-à-vis other systems, and thus time is the primary differential of this paleophysics. This time lock or base time may be defined as the geometric configuration of all entangled and rotating fields at the moment a given system comes into existence, or "comes on line."

- Thus, to harness the energy of space itself, it is necessary to reduplicate the geometric configuration of its significant galactic, solar, and terrestrial systems and physical constants in the practical application of the "as above, so below" principle.

- Since harmonics is the basis of this physics, the means of engineering the "as above, so below"" principle is to engineer local space-time via coupled harmonic oscillators that embody the physical and geometric configuration of the systems from which energy is to be drawn.

(5) Encoded Harmonics of the Planck Units in Ancient Texts

McClain's study of Pythagorean Platonism and our own analysis of it revealed:

- That Plato's works encode a sophisticated system of equal musical tempering;
- That within that system, harmonics is viewed as a means of the unification of the fundamental systems of physics;
- That this unification was achieved by close approximations of the pure theoretical and naturally occuring harmonic series;
- That redundant harmonics of close approximations of the Planck units exist in the Platonic texts, and that other non-western texts may contain similar encoded harmonics of the Planck units.

(6) Encoded Harmonics of the Planck units in the Great Pyramid

To function as a weapon coupling nuclear, electromagnetic, acoustic and gravitational energy together, some harmonic approximations of the Planck units would have to be redundantly incorporated in the structure.

- The inner chambers of the Great Pyramid – the Queen's Chamber, the Grand Gallery, the Antechamber, and the King's Chamber – all contain harmonics of approximations of various Planck units, strongly indicating an engineered unification of quantum mechanics and gravity.
- Since these harmonics are all resonant to each other, and Dunn makes clear that the energy of these chamber is concentrated into the King's Chamber.

(7) Phenomena and Principles of Weaponization Thus Far Suggested

From the previous chapters and the above considerations, some of the principles of weaponizing this form of physics may be postulated:

- Since the laws of this paleophysics were scale invariant, then all levels of physical reality from the quantum to the galactic scale operate according to the principles of the same geometry.
- Since one may draw energy from various entangled systems via coupled harmonic oscillators of great sophistication, the purpose of this coupling would seem to be *to engineer large scale systems such as a target in terms of some quantum state.*
- Put differently, since the aether is a substrate of information in the field and therefore is a non-local reality, the energy of distant systems may be drawn upon via harmonic oscillation simply by reproducing as exact a geometric configuration in the oscillator as possible. This is why the Great Pyramid is constructed not only as an analog of terrestrial and solar physics, but of galactic physics as well.[4]
- Since the primary means of accessing these energies must be via an engineering of Bell's non-locality theorem in coupled harmonic oscillators, the Great Pyramid was a weapon of *non-linear* directed energy, and thus, Dunn's hypothesis that cohered microwave and acoustic output was the primary

---

[4] A means of verification of this hypothesis immediately and happily presents itself, and that is to see whether there are similar nested *redundancies* within the Pyramid that duplicate some harmonic of various physical and geometrical properties of the galaxy itself, such as its mass, its mean mass density, and so on. Further investigations should also include cosmological mass distribution and density, and so on. This has not been done simply because most pyramidologists have not thought to look closely for such things. I predict that such dimensions will indeed be discovered. It is well-known that certain celestial alignments with the Great Pyramid would have required exact geometric knowledge of the galactic equator, and so on.

energy output of the machine is not correct, though is true as far as it goes.

- That being said, it is evident from Dunn's analysis that some effective coupling of acoustic, electromagnetic, and nuclear forces was achieved. The sonoluminesence effect further suggests that the right acoustical interference with a target can induce ionization and nuclear reactions. Cold fusion further corroborates the idea that little understood electromagnetic processes are capable of similar results.

- From the previous considerations, it is evident that there is a missing component of energy that is in all likelihood the primary energy being accessed by the Great Pyramid and utilized as a weapon. Thus we posit that it has a secondary energy output – the *linear* cohered electromagnetic and acoustic component of Dunn – and a *non-linear* component. The Pyramid's "squared circle" properties are well known. Thus we posit that one form of energy utilized in the Great Pyramid is rotational, vorticular energy.

- Eastland's "HAARP" patents further corroborate the idea that the electromagnetic hydrodynamic properties of the earth's atmosphere and magnetosphere can be weaponized *by the same basic technological and scientific principles* for a variety of defensive and offensive purposes, depending on the geometric configuration of the hardware. This view corroborates the Sanskrit vimana texts as well as the unified field view evident in the physics behind the Pyramid.

(8) Missing Components

Sitchin and Dunn, on the basis of entirely different types of evidence, each conclude – apparently independently - that the Great Pyramid is missing its most vital components.

- For Dunn, these missing components consist of banks of Helmholtz resonators, resonant to the various harmonics of the earth, arrayed inside the acoustic amplification chamber, the Grand Gallery.

- Dunn also indicates that some sort of machine to establish vibrations to "prime" the Pyramid might once have existed in the subterranean chamber, and possibly in the underground chambers of the other Pyramids of Giza as well.
- For Sitchin, however, the missing components consist of "ray-emitting crystals" inside the Grand Gallery, giving off a multi-colored rainbow of dazzling light in the chamber. Other "magic stones" determined the "destiny" and "guidance" of the weapon.

B. How it All Worked: The Background Physics, Part One:

Since this unified physics and correspondingly unified technology made use of the above principles, it is no accident that the Pyramid made use of celestial and terrestrial relationships and that these relationships have been misinterpreted by the "time capsule" and "observatory" hypotheses and totally overlooked by the "machine" hypothesis of Dunn. All other hypotheses overlook the only paleographic evidence that clearly specifies a function for the Pyramid, namely, that it was a weapon utilizing these principles. However, it should be noted that the "time capsule" and "observatory" aspects of the Pyramid are essential to its function *as* a weapon.

The weapon hypothesis must therefore explain the basic functions of purposes of:

(1) the celestial relationships embedded within the structure;
(2) the same mathematical relationships found in discrete or separated parts of the Pyramid and how the possible internal relationships between them contributed to its function. As Dunn repeatedly notes, the terrestrial geometries are found in the Pyramid to "increase its efficiency" as a machine. By the same token then, the solar and galactic geometries present in it must be for a similar purpose.

239

(3) The functions of the inner chambers of the Great Pyramid

(4) The possible functions of the other structures at Giza.

Let us now recall the implications of the texts examined in chapter two, and particularly the texts examined by Zechariah Sitchin, in a series of postulates:

These considerations rise to a peculiar answer to the question raised by the "time capsule" hypothesis. Briefly, this hypothesis maintains that the Pyramid was built as a monument to "preserve the knowledge" that the paleoancient Very High Civilization had acquired for posterity. In some versions this hypothesis is maintained because that civilization presumably faced some great cataclysm. In other versions it assumes the form of the "prophecy-in-stone" hypothesis, that its builders possessed some divinely inspired prophetic knowledge of future events that they then encoded in the mathematical dimensions of the structure.

However, since the Pyramid could very easily have been destroyed by any civilization that purportedly possessed nuclear weapons, then question then becomes one of why it was left standing after its internal components were removed upon the order to destroy the weapon, especially since that weapon was so greatly feared. Clearly then, the internal components were crucial to its functioning as a weapon since their removal rendered it non-operational. Presumably such components could, however, have been rebuilt and the weapon again made operational. So why was the structure left standing? This is an important clue. Apparently there was no fear that such components could be easily reconstructed. Two basic explanations present themselves as to why this is so:

- They were too expensive to produce in the aftermath of a major war. But this alternative should be discounted, for the mere potential of their reproduction would seem to be unacceptable to the victors of Sitchin's "Second Pyramid War," since it had been fought for the purpose of destroying the weapon.

- The society that built the Pyramid was considerably more advanced that those of the victors in that war, and its technological infrastructure was decimated to the point that the knowledge and/or infrastructure know longer existed to produce its missing components. In short, the Very High Civilization had begun its long decline to the levels of "civilization" of the ancient worlds of Egypt, Sumer, and the Indus Valley. This would appear to be the more likely explanation.

It is therefore reasonable to assume that it was left standing as a *memorial or monument of that war* and as a warning against the misuses of the technology that enabled its design and construction, much like the Nazi death camps are left as memorials, not only to commemorate the victory over an evil regime, but also to commemorate its victims and the evil uses to which technology was put. On this basis the Pyramid is indeed a "time capsule" containing a message and "lost knowledge" (literally in the form of its missing internal components). However, the message and knowledge that time capsule were meant to preserve have been grossly misinterpreted by current advocates of the "time capsule" hypothesis. The message it was intended to convey after its destruction was a moral one, not a scientific or prophetic one.

Sitchin's texts attribute to this weapon, and particularly to its missing components, the necessary parts of a sophisticated weapons system: a tracking and targeting system, and a "pulsed" beam. Thus if the Pyramid was a weapon, then one should expect a "military" look not only to its but the surrounding architecture. It is to be noted that the Giza plateau does have a peculiar resemblance to the military architecture of a modern military base phased array radar base(cf. figures 1 and 2 below).

Figure One:
Mycerinus, Cephren, Cheops

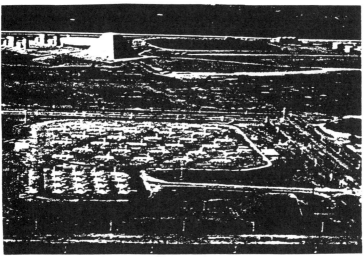

Figure 2:
American Phased Radar Array and Surrounding Military
Installations

The close resemblance of these functional architectures allows at
least one prediction to be made. The high energies involved in such

a weapon would necessarily mean that there would have to be extensive underground bunkers in or near the Giza complex for housing the weapon's crew, the necessary targeting computers, and food, water, and other support bunkers.[5] It is therefore predicted that such chambers do exist and that upon careful examination those chambers will correlate to a military architecture and purpose.

We are now in a position to extrapolate the principles of this weaponized paleophysics upon the following foundations:

- a comparison of the Pyramid's known mathematical and physical properties;
- the relevant ancient texts,
- anomalous phenomena within modern physical science;
- current theory within modern physics;
- an extrapolation of theoretical principles from the Pyramid itself.[6]

Given the implications of this physics for limitless energy as well as its potential for military uses, however, one must also acknowledge the strong possibility that such research may already have been undertaken by institutions or parties interested in

---

[5] Some of the so-called "temples" at Giza display remarkably similar and precise engineering and are of similar antiquity. Could these "temples" at one time been used for such purposes?

[6] This is the most problematical, though nonetheless necessary, component of the methodology. Problematical, in that rather than explaining the Pyramid by the known principles of physics, it departs from a strictly scientific method at this precise juncture. Necessary, however, in that all known and accepted principles of contemporary standard theoretical physics can take one only so far but no farther. To a great extent, the paleographical evidence *demands* such speculation. Since there is abundant evidence that the physics and engineering involved in the Pyramid was *at least* as advanced as our own, we are justified in using the Pyramid itself as a basis from which to extrapolate the principles of a new theoretical construction.

monopolizing and deploying such technology to secure their own power.[7]

The basic principle of the paleophysics of the "Giza Death Star" may now be clearly stated:

Any technological utilization of a theoretically unified physics for weaponized purposes must employ an engineered systems state *harmonically coupled to all the systems to be oscillated in such a way as to amplify all of them,* i.e., it must couple to the "base" or "fundamental" terrestrial system, as well as to the base solar and celestial systems to be oscillated and amplified. This obviously implies that there is a physical and theoretical basis to the astrological doctrine that "destiny" is an effect of the geometric configuration of the heavens (the base solar and celestial systems) in relation to the earth (the base terrestrial system).

These considerations suggest the following physical and terminological model lies behind the Giza Death Star:

(1) The harmonic fundamentals and resonances, that is to say, the cyclotronic and Schumann resonances, of the base and target systems must be reproduced in the coupled oscillator (the weapon itself), where "base system" or "base terrestrial system" or "base planetary system" are synonymous, and defined as the inertial frame of reference for a planetary observer – i.e., the weapon itself - in that system;

(2) The coupling of the oscillator to the base or fundamental planetary system may or may not have to occur by actual physical contact with the base system, though this is the case with the Pyramid itself.[8] This suggests that the key to such technology lies not along digital, but along analog,

---

[7] It is reported that the Schorr expedition's results have been sealed by the Egyptian government, which may or may not have been acting on its own, or which may or may not have been the ultimate source of the sealing order.

[8] That is to say, at the current stage of research, one cannot say with certainty what the principle of the paleophysics was, i.e., whether or not it *required* such local physical contact with the base planetary system. My intuition, however, tells me that it did.

harmonic coupling. Any digital element would have been primarily utilized in the design, information transcription, and "tuning" (targeting) of the weapon. Since the Pyramid and the Giza complex as a whole are also analogically coupled to the base solar and celestial systems, this suggests that non-local systems may be harmonically coupled by analogs of those systems. This in turn implies that the basis of such coupling is on some practical application of Bell's non-locality theorem in order to transfer the energies of those systems at any given moment;

(3) The coupled oscillator-amplifier (the Pyramid itself) must employ several "nested" layers of these harmonic relationships to amplify the harmonics of the geometry of the total acoustic, inertial, and electromagnetic energy of the base planetary, solar, and celestial systems. Moreover, the careful placement of those dimensions within the structure will *direct* the various forms of energy to the proper place for coupling to each other. This would have *required* computer-aided designing;

(4) The primary acoustic energy input of the coupled oscillator-amplifier derives from the base terrestrial system;

(5) The coupled oscillator-amplifier must have an acoustically and optically resonant amplification chamber resonant to the base planetary system's Schumann resonance (Grand Gallery);

(6) It must likewise be resonant to the thermal and mass gradients of the base terrestrial system;

(7) The tracking and targeting of the weapon occurs as a result of a harmonic interferometry, which indicates one possible function of the other Giza structures as well as the possible function of some of the Pyramid's missing components;

(8) Thus, the Great Pyramid, and the Giza complex as a whole, gives a comprehensive – though due to the absence of some internal components, not a complete - list of the required mathematical and physical properties of such a weapon.

The use of such a weapon, while far more horrific than the use of thermonuclear weapons in the targeted area, would not necessarily involve collateral damage to the society possessing and deploying it. However, theoretically, such a weapon would be capable of a "one shot planetary destruction." Some means of calibrating and damping the energy output of the weapon would therefore also have to be incorporated in the original design.

(1) Definitions:

These principles require some definition of the types of systems utilized in drawing and directing the energy of the weapon.

(a) of Systems:

Base Planetary System, Fundamental System, and Base Terrestrial System: These terms are synonymous and define the planetary system to be harmonically oscillated and amplified, in this case, the earth. It is the inertial frame of reference of the oscillator - weapon, which is in turn *defined as the observer.*

Base Solar, or Base Stellar, System: These terms are synonymous and define the solar system, including the stellar center of that system, in which the base planetary system exists. It is inclusive of the stellar mass, the total planetary mass, and any and all possible geometric configurations of the elements of that system, including any significant planetary bodies possessing angular momentum relative to its stellar center.

Base Celestial, or Base Galactic, System: These phrases are synonymous and indicates the galactic system in which the base planetary and solar systems exist and possess angular momentum relative to the galactic center of mass.

It is evident from these considerations that the primary form of energy being accessed by the Great Pyramid is *inertial*, i.e., the very energy that results *from a particular configuration of space.*

Accordingly, its builders also had to presuppose that the energy of a non-local system, for example, the galactic center, was transmitted via the superluminal transcription of information in the aether hyperspace. They are relying, in other words, on some paleophysical version of Bohm's notion of a "pilot wave" and Bell's non-locality theorem to draw energy from these distant sources.

From these considerations it is further evident that any fundamental planetary system to be oscillated and amplified must be harmonically coupled to certain fundamental physical geometries:

(1) the precession of the base planetary system's equinoxes;
(2) the precession of the base stellar system's equinoxes; and to
(3) the base celestial system's equator.

In other words, the paleoancient Very High Civilization that built the Great Pyramid was at the minimum a Type II civilization.[9]

The purpose of the analogical or harmonic coupling to these geometries is clear, and has nothing whatsoever to do with any metaphysical or religious explanations subsequently fastened by Egyptian society upon them. The coupling was solely to access the enormous energy latent in these physical configurations via Bell's principle of non-locality. The energy potential of the total system is thus dependent upon the information in the field, that is to say, the geometric configuration of these interlocked systems at any given moment. And thus one may draw a final corollary by means of the verification of the theoretical construct: *the energy yield of any nuclear or thermonuclear device is dependent upon the geometry, that is, the harmonics, of any given time and place that it is utilized.*[10] The geometric configuration of systems is thus the

---

[9] Cf. Chapter three and chapter nine, pp. , .

[10] This may indeed constitute one of the remaining and most closely held military secrets of the nuclear powers.

only measure of time, and time is not therefore a dimension of physics but its primary differential.

(b) Of Base Time, or the Primary Differential

The Base Time of the Total System: If the "total system" is defined as the base planetary, solar, and celestial systems, as well as the coupled harmonic oscillator and amplifier itself locally grounded to the base planetary system (i.e., the Pyramid itself), then the "base time" of the total system denotes the geometric configuration of the base planetary, solar, and celestial systems *at the moment the coupled oscillator and amplifier becomes fully operational.*

It is to be noted that this putative principle of the paleophysics of the Pyramid also requires careful design modeling of the weapon before the construction was actually begun, so that the oscillator would begin to function fully at the precise moment when the chosen geometric configurations of those systems aligned with the oscillator and allowed it to be primed to peak efficiency. Thus not only did the structure have to be computer-designed, so did the timing and process of construction, since the purpose of the device itself is precisely to engineer local space-time for destructive purposes.

Artificial Harmonic Entanglement: Once the locally grounded oscillator and amplifier is so coupled to the three systems and fully functional, the systems are said to be artificially harmonically entangled and will continue to be entangled so long as the geometric analogy of the oscillator-interferometer is not significantly altered. As long as it is an analog of those systems, the systems remain entangled in and by the oscillator.

This is to say that as long as the Great Pyramid stands, notwithstanding its missing components, it will oscillate and draw energy from the earth, a phenomenon amply documented.

* * *

The entanglement of systems via the harmonic (geometric) analog of an oscillator is an essential component of systems – and especially of biological systems – not yet adequately comprehended by theoretical physics. This may explain why attempts to repeat experiments in endothermic fusion or zero-point energy powerplants sometimes fail. T.E. Bearden comments as follows:

> For example, Frank Gordon once developed an electric motor which (Sic.) after several years produced an output power which was 1.67 times greater the input power. What we did not know was that his years of operation and struggle with the motor and its structure had gradually deterministically charged-up (activated) the local atomic nuclei and activated the local vacuum potential. Thus the motor gradually "grew" its increase in power, seemingly violating the conservation of energy *if one tried to regard the motor as a closed system.* (Of course, it was now an <u>open</u> system, receiving input from the structured vacuum as a... source. In other words, Frank had formed a <u>structured quatum potential</u> in the local area, coupled to the machine.)[11]

These are the exact principles in operation at Giza, since the mathematical relationships of the Giza complex and its entangled systems *also* structure the local base planetary potential of the vacuum by drawing on the already structured potential of all three systems.

(c) Technological and Physical Principles Extrapolated from the Giza Death Star

(1) *Piezo-Electric Core:* The oscillator must be *of sufficient mass* and constructed of material able to take advantage of the peizo-electric effect. Sufficient mass, coupled with resonance harmonic oscillation to the Schumann resonance, will not only

---

[11] T.E. Bearden, *Gravitobiology* (Tesla Book Co.). Italicized emphasis mine, underlined emphasis Bearden's.

stress this peizo-electric core but do so in phase with the impulses of that resonance. Its mass must therefore also couple to the base planetary system's electromagnetic and mass gradients in some ratio or harmonic of them.[12]

(2) *Coupling to the Hydrodynamic and Thermal Gradients:* The coupled oscillator must couple in a 1:1 ratio to the mean thermal hydro-dynamic gradient of the base planetary system.

(3) *Base Planetary System Geographic Alignment:* The harmonic oscillator and amplifier should be aligned to the true north-south axis of the base planetary system.[13] Its faces should be parabolic, aligned to the four cardinal compass points of the base planetary system. This is an essential feature of the coupling to the structured potential of the geometric configuration of the various terrestrial, solar, and galactic system.

(4) *Multi-leveled Systemic Coupling within the Oscillator:* The oscillator must "nest" the mathematics of the total system, i.e., must be so constructed that its dimensions themselves are *ratios of the geometry of the total system. The dimensions of the oscillator are themselves harmonic oscillators within the oscillator, resonant to some aspect of the total system.*

(5) *Coupling to Other Primary Base Planetary System Gradients:* the actual resonance chamber (the King's Chamber), baffle chamber (Antechamber), amplifier (Grand Gallery) must couple the acoustic, electromagnetic, thermal, and mass gradients of the base planetary system to each other and to:
(a) the parabolic reflector faces of the oscillator, and to

---

[12] Cf. in this and what follows, careful consideration of the properties of the Pyramid outlined in chapter five are required.

[13] i.e., understood as a spherical rotating mass. It has long been recognized that the Pyramid embodies two spatial geometries in one object, a spherical one and the embedded pyramid itself. Many authors speak of the Pyramid as "squaring the circle." In terms of its machine function, it would be more appropriate to speak of embedding a platonic solid in a sphere.

(b) the "apex".[14]

(6) *Coupling to the Angular Momentum of the Base Solar System:* The actual resonance chamber (the King's Chamber), baffle chamber (Antechamber), and amplifier (Grand Gallery) must couple to the base time and angular velocity of the base planetary and solar systems.

(7) *Coupling to the Base Stellar and galactic Systems:* Said chambers must also couple to the base stellar and galactic systems via the precession of the equinoxes, as must the structure in which they are housed.

(8) *Coupling to π, φ and Other Physical and Mathematical Constants:* The oscillator should couple to physical constants,[15] and particularly to the Fibonacci series of ϕ, with the theoretical value of 1.161818... and with $\pi = 6/5\ \phi$. The purpose of this coupling is *perhaps* to access the EPR effect and utilize Bell's theorem for the transcription of information in the field of the total system, i.e., the energy of the total system. Other such constants would be 2.72, $\sqrt{2}$, $\sqrt{5}$, $\sqrt{6}$, $\sqrt{7}$ and so on.

(9) *Coupling to Two or More Spatial Geometries:* The coupled oscillator should couple two or more of the spatial geometries of the Platonic solids.[16]

(10) *Coupling to the Base Solar and Celestial Systems' Linear Velocity of Light (c):* The coupled oscillator should express

---

[14] If one calculates the dimensions of the missing Apex stone from the known dimensions of the Pyramid, it would appear that the missing capstone is almost exactly 1/100th the size of the whole structure.

[15] Here is yet another means of verification of the hypothesis. If the Pyramid was constructed by a paleoancient Very High Civilization that possessed the type of unified physics and technology being summarized here, then it is inconceivable, in so structuring local vacuum potential, that such a civilization would have been ignorant of the fundamental constant of quantum of action, or Planck's constant, *h*. To my knowledge, no one has yet looked for, nor found this constant in the Great Pyramid because no one has thought to look for it there. If it *is* found, it will constitute one of the most startling confirmations of the sophistication of the society that built it.

[16] Cf. the Appendix.

these systems' linear measure of the velocity of light as a dimension, or dimensions, of the oscillator in some ratio to the mean distance from the solar center to the base planetary system.

(11) *Coupling to the Radius/Diameter of the Base Solar System's Stellar Source:* The oscillator should couple to the measure of the base solar systems radius by some ratio present in the dimensions of the structure.

(12) *Coupling to the Magnetic Poles of the Base Planetary System:* The coupled oscillator should couple to the angular momentum of the base system's electromagnetic field.

(13) *Coupling to the Gaussian Constant of Gravitation:* The harmonic oscillator must couple to the Gaussian constant of gravitation of the base planetary and solar systems. This strongly suggests that at the heart of the resonator-amplifier-baffle array (i.e., the King's Chamber, Grand Gallery, and Antechamber respectively), a gravitational phenomenon or effect is being manipulated, i.e., that local-space time is being engineered.

(14) *Coupling to the Spatio-temporal Dimensions of the Base Planetary System:* The oscillator must couple to the diameter, circumference, volume, and mass of the base planetary system.

(15) *Coupling to the Electromagnetic Cavity of the Base Planetary System:* The oscillator must couple, in one or more of the dimensions present in the structure, to the Schumann resonance of the Base Planetary System, i.e., digitally as well as analogically.

(16) *Coherence of All Acoustic and Electrodynamic Input into the Oscillator.*[17]

(17) *Entangle the Cohered Microwave Output with the Acoustical Harmonic information of the Oscillator and its Entangled systems:*

(a) The purpose of this principle is speculative in nature. The weaponized harmonic interferometry being posited as the basis of the Giza Death Star is that the microwave and acoustic output act as carrier wave to induce a longitudinal wave in the target, causing it to cavitate and undergo a kind of combined sonic-electromagnetic disruption.

(b) Thus also suggests another putative principle of paleophysics. The nuclear reactions possibly driving or "powering" the oscillator[17] may exist in a plasma state that separates protons and electrons.

Again, these are but the extrapolations of the principles of physics and technology that the Pyramid itself suggests.

Now we must pause a moment to say a few more words about Christopher Dunn's ingenious power plant hypothesis once again. Suppose that the paleoancient Very high Civilization really did build the Pyramid as some sort of power plant to provide wireless electrical power, along similar lines that Nichola Tesla once indicated that he could provide wireless electrical power all over the face of the earth with a very few "transmitters." Indeed, Tesla also indicated that the same technology could deliver a "death ray" of enormous destructive power to a target. To do so would require but a simple "retuning" – to use Tesla's word - of the power plant to focus its electrical transmission, from a diffuse earth-enshrouding field to a concentrated bolt pinpointed on a target. In other words, the unified physics being postulated for the paleoancient Very High Civilization allows the same device to function in *both* roles by simple reconfiguration of the system. However, certain aspects of the power plant version of the machine hypothesis now fail, and fail in the light of Mr. Dunn's own analysis. For example:

(1) He assumes the existence of "extra-terrestrials" that would require a microwave signal output. In other words, the microwaves are cohered and enormous power output is required for interstellar communication. The problem is that the suggested physics embodied in the Pyramid *exceeds* our contemporary principles of physics, and this would in turn seem to make such slow methods of

---

[17] Cf. the discussion of the use of hydrogen in the previous chapter.

communication between star systems rather inefficient. So the purpose of the microwave output of the Pyramid does not appear to be for the purposes of communication primarily. Why put such extraordinary effort and expense into an interstellar long distance phone call, since the whole principle of non-locality would have made microwave communications as efficient to that society as the Pony Express is to our own?

(2) Dunn also emphasizes the piezo-electric effect is being utilized for electrical output. But again, the principles of physics embodied in the Great Pyramid suggest a physics beyond our own. If Tesla could claim to transmit wireless power *without* millions of tons of granite, then the granite that constitutes the overwhelming bulk of the Pyramid's must have served some other function that required such an enormous mass.[18] In other words, if one accepts, as both Dunn and this author do, the hypothesis of a technologically sophisticated very ancient culture, then the biggest anomaly of all still remains. Why build such an expensive and enormous pile of rocks just to have a maser and electricity, or just to "talk to the stars" in some paleoancient SETI project? The anomaly of the size, purpose, and expense still remains. Military projects, however, are well-known for both size and expense, and human nature being what it is, I see no reason why the same should not have been true in ancient times.

(3) Finally, there is the question of the Grand Gallery – Dunn's resonance chamber – itself. What power plant employs banks of Helmholtz resonators in an amplification

---

[18] Another point must be stressed here, and that is the *expense* of building such a structure. Presumably a paleoancient Very high Civilization would have had the technology to build power plants without the expense of quarrying and moving tons of granite, so some *functional purpose* is left unexplained, a purpose justifying the enormous outlay of financial resources to construct such a labor-intensive structure. The only analog to such expenses in our own culture is sophisticated military systems.

chamber? Again, presumably for signals to modulate a microwave carrier wave, which is more appropriate to a communications function. But then, why cohere that output in a maser (the Coffer)? If a society had the capability for interstellar travel that Dunn suggests it had, then it had other more direct means of communication between planetary systems. Why build an enormous pile of precisely tooled granite, weighing millions of tons, align it precisely to every conceivable principle of celestial and terrestrial mechanics, and fill it with Helmholtz resonators, just to "phone home?" Clearly something more than just "power output" or communications would seem to be implied, given the enormous energies that are being coupled.

In short, Dunn's bold and ingenious explanation is indeed brilliant, but raises as many questions as it answers since he limits himself to explanations guided by current paradigms of orthodox science and technology. In fairness to him, it should be noted that one senses much perplexity on his part in the final pages of his book as he wrestles precisely with these types of questions.

Of course, use of the putative principles of paleophysics to build power plants *is* possible, but in that case, one would expect several much smaller, less expensive installations, perhaps one in each paleoancient household, drawing energy from the heavens. The enormity of the Giza Death Star, and the principles it embodies, were, as Sitchin's texts indicate, for but one purpose.

C. How it all Worked: The Background Physics, Part Two: Tesla's High Frequency Direct Current "Impulse" Technology

The vast influence of the electrical work of Nichola Tesla is known to most informed people. Less well-known are the directions that Tesla's experimental research took from the close of the 19th century to the end of his life. Stories bordering on the mythological surround this period, stories of mysterious forces, and of government agents scurrying to confiscate his papers and notes upon his death. Whether those stories are true – and I am inclined to believe that they are – the work of the last period of his life, incorporating some of Tesla's most brilliant experimental insights, affords a crucial look into the nature of physical reality as well as an understanding of the conflict between physics as an experimental science and as a theoretical and mathematical discipline. Moreover, Tesla's late work, and its subsequent profound misinterpretation by theoretical physics' orthodoxy, demonstrate the degree to which the received theories and paradigms of "normal science" can inhibit scientific insight, and be manipulated by vested power elites to close down lines of inquiry threatening to their own basis of power.

The author and science researcher Gerry Vassilatos has long investigated the "forgotten" highways of the physical sciences, and of the peculiarities of Tesla's last lines of research and his own extraordinary claims with regard to its positive and negative potentialities. Vassilatos' work, *Secrets of Cole War Technology: Project HAARP and Beyond*, contains the clearest account in public literature of the "electrical impulse" investigations that so consumed Tesla in his later life. In this section we rely upon Vassilatos' account to summarize the experiment that led Tesla to investigate a whole new electromagnetic phenomenon, as well as the subsequent experiments he devised to confirm and expand his knowledge of it. These were precisely the experiments upon which Tesla based so many of his extravagant claims for a new source of limitless energy, as well as his seemingly fantastic claims for a weapon of mass destruction of planetary-busting power.

Vassilatos begins his account as follows:

> But while endeavoring toward his own means for identifying electrical waves, Tesla was blessed with an accidental observation which forever changed the course of his experimental investigations.... Part of this apparatus (was) ... a very powerful capacitor bank. This capacitor "battery" was charged to very high voltages, and subsequently discharged through short copper bus-bars. The explosive bursts thus obtained produced several coincident phenomena which (sic., et

passim) deeply impressed Tesla, far exceeding the power of any electrical display he had ever seen. These proved to hold an essential secret which he was determined to uncover.

The abrupt sparks, which he termed 'disruptive discharges', were found capable of exploding wires into vapor. They propelled the very sharp shockwaves which struck him with great force across the whole front of his body. Of this surprising physical effect, Tesla was exceedingly intrigued. Rather like gunshots of extraordinary power than electrical sparks, Tesla was completely absorbed in this new study. Electrical impulses produced effects commonly associated only with lightening. The explosive effects reminded him of similar occurrences observed with high voltage DC generators. A familiar experience among workers and engineers, the simple closing of a switch on a high voltage dynamo often brought a stinging shock, the assumed result of residual static charging.[19]

This phenomenon led both power company engineers and Tesla to speculate on the reasons for this strange discharge. It should also be noted that the effect Tesla was obtaining bears some resemblance to the "electro-hydro-dynamic" phenomena being observed by Hannes Alfvén mentioned in chapter three.

The theoretical and metaphorical framework in which Tesla framed his explanatory hypothesis for his next series of experiments points to a profound and persistent problem in theoretical physics, from relativity to quantum mechanics. At this juncture, it is important to recall that Tesla formulated his explanation *before* either of these theoretical bulwarks of modern physics were formulated.

Tesla knew that the strange supercharging effect was only observed at the very instant in which dynamos were applied to wire lines, just as in his explosive capacitor discharges. Though the two instances were completely different, they both produced the very same effects. The instantaneous surge supplied by dynamos briefly appeared superconcentrated in long (power) lines. Tesla calculated that this electrostatic concentration was several orders of magnitude greater than any voltage which (sic., et passim) the dynamo could supply. The

[19] Gerry Vassilatos, *Secrets of Cold War Technology: Project HAARP and Beyond* (Bayside, California: Borderland Sciences: 1996), p. 26.

actual supply was somehow being amplified or transformed. But how?[20]

    The general consensus among engineers was that this was an electrostatic 'choking' effect.... Like slapping water with a rapid hand, the surface seemed solid. So also it was with the electrical force, charges meeting up against a seemingly solid wall. But the effect lasted only as long as the impact. Until current carriers had actually 'caught up' with the applied electrical field, the charges sprang from the line in all directions.... (Tesla) began wondering why *it was possible for electrostatic fields to move more quickly than the actual charges themselves, a perplexing mystery.*[21]

That is, Tesla knew that the electrical current moved at approximately the speed of light. But this in turn meant that the electrostatic field itself was moving at a superluminal velocity.

The careful reader will now recall that this is almost exactly the same position as quantum physicist David Bohm's supposition of a superluminal "pilot wave" guiding the slower luminal velocity electrons along their paths.[22] As Tesla saw it, the problem was that the brief, almost instantaneous application or "impulse" of electrical power impacting against the resistance barrier brought on an abnormally "electro-densified condition."[23] Through experimentation, Tesla determined "that he could literally shape the resultant discharge, *by modifying circuit parameters. Time, force, and resistance were (the) variables necessary to producing (sic.) the phenomenon.*"[24]

At this juncture, recall that Alfvén likewise maintains more or less the same thing, time, force, and resistance were variables that appeared to follow laws that were scale invariant from laboratory sized experiments to galactic superclusters. Note secondly that Tesla also makes the discovery that the geometric configuration of

---

[20] Tesla's mystification would be that of any competent physicist, for the increase of energy would appear to violate the Second Law of Thermodynamics.

[21] Vassilatos, op. cit., p. 27.

[22] Cf. Chapter Three.

[23] Vassilatos, op. cit., p. 28.

[24] Ibid.

circuit parameters it itself a factor determining how much energy was released in the discharge.

Finally, an analogy may be helpful to explain the phenomenon. Tesla's reasoning behind the discharge was essentially this. At the very moment that the electrons of the current spark struck the wire or bus-bar, the geometry and density of the atoms in the bar effectively raised the resistance of the wire to infinity. No matter how high the current of the spark, the discharge still occurred. Electrons hit the resistance barrier, and splattered out from the surface of the bar in all directions perpendicularly. We're all familiar with another form of this phenomenon. We've all climbed a diving board and jumped into a swimming pool, only to "belly-flop" and smack against the resistance of the surface of the water. At that very instant, no matter how fast we jump or how much we weigh, we hit up against that momentarily "infinite" resistance of the water's surface, and send water splashing out in all directions around us. What Tesla did was to break the current at the very moment the electrons hit the surface of the wire, much like if we could, at the very instant we hit the water, "run the film backwards" and do it over and over again in quick succession. With this simple analogy in mind, let us continue with Tesla.

In order to test the phenomenon further, Tesla resolved to repeat the experiments with direct current to eliminate the "backrush" to the dynamo caused by alternating current.

The result this time was even more astonishing:

> The sudden quick closure of the switch now brought a penetrating shock wave throughout the laboratory, one which could be felt both as a sharp pressure and a penetrating electrical irritation. A 'sting'. Face and hands were especially sensitive to the explosive shockwaves, which also produced a curious 'stinging' effect at close range. Tesla believed that material particles approaching the vapor state were literally thrust out of the wires in all directions.[25] In order to better study (sic.) these effects, he poised himself behind a glass shield and

---

[25] What Tesla meant by "the vapor state" would be approximately what a quantum physicist would mean by quantum or sub-quantum particles, what we have called alternatively "quantumstuff" or aether.

resumed the study. Despite the shield, both shockwaves and stinging effects were felt by the now mystified Tesla. This anomaly provoked a curiosity of the very deepest kind, for such a thing was never before observed. More powerful and penetrating than the mere electrostatic charging of metals, this phenomenon literally propelled high voltage out into the surrounding space where it was felt as a stinging sensation.[26]

Throughout these experiments, in other words, Tesla was not only observing over-unity energy output – getting more energy *out* than he was putting *in* - but feeling shockwaves *apparently oblivious to the normal shielding effects of matter.* He was getting more energy out of the system than he was putting into it, and feeling waves that traveled clean through solid objects like they were so much air. No wonder he was mystified! But he made the appropriate conclusion; *his system was not a closed system, but an open system, and he was somehow accessing a source of energy outside the system by dint of some inherent properties of the configuration of the system itself.*

In 1892, Tesla published a lecture in which he detailed these experiments. Titled "The Dissipation of Electricity," this lecture marks the point in Tesla's career where he abandoned research into high frequency alternating current for good in order to conduct a new line of experiments to describe the phenomenon of high energy direct current impulses and the resulting "shockwaves."[27]

> He now prepared an extensive series of tests in order to determine the true cause and nature of these shocking air pulses. In his article, Tesla describes the shield-permeating shocks as 'soundwaves (sic.) of electrified air.' Nevertheless, he makes a remarkable statement concerning the sound, heat, light, pressure, and shock which he sensed passing directly through copper plates. Collectively, they 'imply the presence of a medium of gaseous structure, that is, one consisting of independent carriers capable of free motion.' Since air was obviously not this 'medium', to what then was he referring? Further in the article he clearly states that 'besides the air, another medium is present.'

---

[26] Ibid., p. 29.
[27] Vassilatos, op. cit., p. 31.

Through successive experimental arrangements, Tesla discovered several facts concerning the production of this effect. First, the cause was undoubtedly found in the abruptness of the charging. It was in the switch closure, the very instant of 'closure and break,' which thrust the effect out into space. The effect was definitely related to time, *impulse* time. Second, Tesla found that it was imperative that the charging process occurred in a single impulse. No reversal of current was permissible, else the effect would not manifest. In this, Tesla made succinct remarks describing the role of capacity in the spark-radiative circuit. He found that the effect was powerfully strengthened by placing a capacity between the disruptor and the dynamo. While providing a tremendous power to the effect, the dielectric of the capacitor also served to protect the dynamo windings.... The effect could also be greatly intensified to new and more powerful levels by raising the voltage, quickening the switch 'make-break' rate, and shortening the actual time of switch closure.[28]

It is now to be noted that the granite core of the Great Pyramid would function like a giant capacitor in Tesla's experiment, since its piezo-electric properties, under constant stress both from the mass of the structure as well as its resonance to the Schumann resonance would so stress the core as to build up a phenomenal charge. *In other words, the Pyramid utilized some form of the same impulse energy discovered by Tesla.*

Tesla's experiments also reveal more properties relevant to the Pyramid's function as a weapon. He discovered that it was possible to amplify the shockwave effect of the impulse by an *asymmetrical geometric arrangement of the system's components.*[29] "By placing the magnetic discharger closer to one or the other side of the discharging dynamo, either force positive or force negative vectors could be selected and projected."[30] That is, the power output of the system varied as a function of the geometrical configuration of its components, exactly what our model of the hyperdimensional physics predicts. Precisely such asymmetry is found in the internal

---

[28] Vassilatos, op. cit., pp. 31-32, emphasis in the original.
[29] Ibid., p. 34.
[30] Ibid.

chambers of the Great Pyramid, as well as the asymmetrical arrangement of the other structures at Giza.

More importantly, "Tesla found it impossible to measure a diminution in radiant force at several hundred yards. In comparison, he recalled that Hertz found it relatively easy to measure notable inverse square diminutions.... Tesla suspected that these effects were coherent, not subject to inverse laws other than those due to ray divergence."[31] We have already discovered that Dunn convincingly argues that a cohered microwave output was involved in the Great Pyramid. But Tesla's discovery interests us for another reason, and that is the reliance of his open system upon geometric configuration to tap some unknown source of energy not subject to normal inverse square relations. And that energy was some sort of electro-acoustic *cohered* longitudinal wave. It remains to be shown how cohered microwaves and electrical impulses might be connected and might have been weaponized in the Great Pyramid.

But Tesla's most astonishing hypothetical model concerns the actual over-unity energies he claimed to observe in his experiments.

> Actual calculation of these discharge ratios proved impossible. Implementing the standard magneto-inductive transformer rule, Tesla was unable to account for the enormous voltage multiplication effect. Conventional relationships failing, Tesla hypothesized that the effect was due entirely to radiant transformation rules, obviously requiring empirical demonstration. *Subsequent measurements of discharge lengths and helix attributes provided the necessary new mathematical relationship.*
>
> He had discovered a new induction law, one whose radiant shockwaves actually *auto-intensified when encountering segmented objects. The segmentation was the key to intensifying the action.* Radiant shockwaves encountered an helix (sic.) and 'flashed over' the outer skin, from end to end. This shockwave did not pass through the windings of the coil at all, treating the coil surface as an aerodynamic plane. The shockwave pulse auto-intensified exactly as gas pressures continually increase through Venturi tubes. A consistent increase in

---

[31] Ibid.

electrical pressure was measured along the coil surface.... Tesla further discovered that the output voltages were mathematically related to *the resistance of turns in the helix. This resistance meant higher voltage maxima.*[32]

The importance of Tesla's observations and of their application to the Pyramid cannot be overestimated. Let us see how they closely correlate to what is found in the Great Pyramid.

Viewed in one sense, the granite core of the Pyramid, its vast stone courses and the geometric shape of the Pyramid itself as a "squared circle" would function as a capacitor reliant upon the piezo-electric properties of the granite itself. However, viewed in yet another sense, The Pyramid, precisely as a squared circle, is *an electrical coil that is segmented – exactly in accordance with the principles discovered by Tesla – not only into separate "windings" in the stone courses, but each of these "windings" in turn is segmented into a discrete number of stones.*[33] The pyramidal form itself gives the distinctive geometry and properties of a Tesla impulse coil (cf. figure 3).

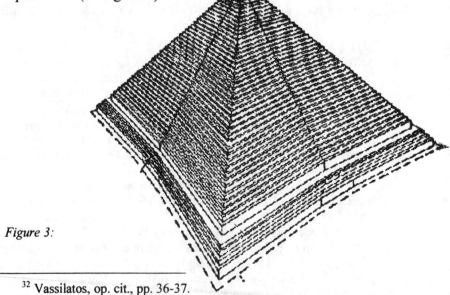

Figure 3:

---

[32] Vassilatos, op. cit., pp. 36-37.

[33] And like everything else in the Pyramid, the exact number of stones in each course or winding is probably the result of an exact mathematical and physical design.

The actual "pulses" of the Pyramid – the "switch closure" durations –are resonant to the Schumann resonance.

Another important distinction is to be made between this genuine Tesla impulse coil and the coils commonly called Tesla coils and found in so many high schools and college physics laboratories, but having nothing to do with this sort of electrical impulse technology.

> Tesla transformers are not magnetoelectric devices, they use radiant shockwaves, and produce pure voltage without current. No university high frequency coil must ever be called a 'Tesla coil', since the devices usually employed in demonstration halls are the direct result of apparatus perfected by Sir Oliver Lodge and not by Nikola Tesla.... Tesla transformers produce extraordinary white impulse discharges of extreme lengths and pressure, which exceed the alternating violet spark displays of Lodge coils.[34]

We are now in a position to hypothesize a possible physical and mathematical model of this impulse technology and its possible utilization in the Pyramid. In this hypothetical interpretation, observe how each aspect of the Tesla impulse technology finds an analog in the basic structure of the Pyramid.

- The granite stone courses function as the coil windings of the system, their resistance multiplying the voltage of the impulse output;
- The granite stone blocks function as the segments of the system, multiplying the voltage of the impulse output;
- The shape or geometric configuration of the system is a coil (a "squared circle" much like a Platonic solid) in a harmonic relationship to the system or systems to be pulsed or targeted;
- The mass of the system is the mass of the granite, which is in turn a harmonic of the mass of the base planetary system;
- This mass stresses the piezo-electric properties of the granite, pulsing in resonance to the Schumann resonance of the earth

---

[34] Vassilatos, op. cit., p. 37.

and simultaneously functioning as a capacitor building up charge when not in "use."

On this model, a number of pulses would be directed by a harmonic interferometry to target, *which might indicate the possible function of the other pyramids or some of the missing internal components of the Great Pyramid itself.* When being discharged or "firing," the Pyramid would possibly have been sheathed in a momentary pale blue light, a plasma moving from the base of the structure to its apex, where it would simply have disappeared, reappearing as a bolt of lightening on the target, as Tesla's "electrified air" shockwave, inducing an acoustic cavitation in the nuclei of the target causing a nuclear "meltdown" even in ordinary stable elements until the force of the "acoustical shockwave" was dissipated.[35]

But there are other peculiarities of the Great Pyramid that Tesla's impulse technology, taken alone, cannot explain.

D. Bell's Non-Locality Theorem, and Zero-Point Energy Engineering

When one turns on a radio, one is listening to two types of energy that have become entangled: (1) a carrier wave, the radio wave itself, and (2) acoustic information encoded or modulated into the carrier wave. A radio transmitter combines the two types of information and a receiver disentangles them.

How then would the Pyramid have directed such energy to a target? Recall that Bell's theorem states that at the deepest level, reality is *non-local,* and that Bohm's version of quantum mechanics posits the existence in non-local reality of a "pilot wave" transcribing information to an electron at superluminal speeds.

---

[35] I have not experimented with Tesla's impulse technology, though am aware that the engineer and inventor Eric Dollard has done so.

Thus it seems reasonable to conclude that the builders of the Great Pyramid had to have known about, and moreover knew how to engineer, some version of Bell's non-locality theorem and Bohm's "pilot wave", because the utilization of the inertial energies of the solar system, and more importantly, the galactic system *could be accessed in no other way*. If those systems were harmonically coupled in the Pyramid, as all the physical evidence suggests, then the information in those systems could only be accessed at superluminal speeds. Non-locality is the very basis on which those energies were not only accessed via coupled harmonic oscillation but directed to target. So in this sense the Pyramid is a great collector (hence the parabolic faces?) of this information, as well as a phase conjugator of that information.

The energy of those systems does not "travel" to the Pyramid as much as the information of that energy is transcribed to it. In this sense, it is a "mirror" or "receiver". But additionally, utilizing the same principle of non-locality, the same "aether wave" or "pilot wave" most likely was used *as a carrier wave* for the acoustic, microwave, and electrical impulse output of the Pyramid. For this reason, one must not imagine that when the Pyramid was "firing" that bolts of energy beamed up into space, and bounced off satellites or curled around the earth's magnetic field lines to target. Rather, the bolts of energy, if visible at all, rotated in a vortex upward over the structure, disappeared, and reappeared over the target, or better, *inside* the target. And how were these discharges directed? Harmonically, of course, by tuning or geometrically configuring the missing components – perhaps in conjunction with other structures of the compound - to guide the pilot wave to the target. Theoretically, then, *any* target in nearby space could have been selected, from the Indian sub-continent, to the moon, or Mars.

E. The Sonoluminesence Effect.

In chapter four we saw what the sonoluminesence effect was. Something like that effect is being produced by this weapon. The aether or pilot wave, and the electromagnetic wave are being used both to carry the acoustic wave – the harmonic information – to the target and to accelerate it once inside the target. Each wave, in effect, is acting as its own coupled harmonic oscillator, in phase with the other waves and once they are inside the target, with the target itself. Since the target has no adequate means of damping, it simply vibrates apart, or blows up, at the atomic level, inside out.

F. What Were the Great Pyramid's Missing Components?

(1) The Gravito-Acoustic Oscillator and Amplification Chamber (The Grand Gallery): Its Function and Missing Components

So what might the Great Pyramid's missing components have been? What was, in fact, once inside the Grand Gallery? I believe that Dunn is essentially correct. They were some sort of acoustic resonators, arrayed in banks that fit into the slots on the side ramps. *But* as Sitchin's texts indicate, they were also much more than these. I believe they were artificial crystals whose crystalline structure *as well as their overall geometric configuration* were carefully – and at great expense – engineered to be both optically and acoustically resonant to the three systems – terrestrial, solar, and galactic – that the Pyramid was coupled to. The Mitchell-Hedges crystal skull is evidence that there once existed a very sophisticated knowledge of crystals.[36] This solution seems to fit the evidence adduced by Dunn and Sitchin.

But this may seem rather fanciful. What could the function of such crystals possibly have been? The solution is simple, but breathtaking: *they may have been to resonate to the acoustic harmonics of gravity itself*:

---

[36] And I can now state why I always said qualified the word "carved" when I referred to it in chapter one, since I believe it was not carved at all, at least, not in any standard sense of bulk technology. It was more likely *grown* from the inside out by some process of nanotechnology. Such a technology would have allowed the construction of the types of acoustically and optically resonant crystals I am talking about here.

At Moscow University, Vladimir Braginski is looking for gravity waves by monitoring tiny changes in the shape of a 200-pound sapphire crystal *cylinder. Braginski chose this exotic material because after being struck it continues to quiver for a record time.* Sapphire's long ringing time permits making a maximum number of measurements before the gravity wave impact fades away. To isolate it from terrestrial noise, the Soviet sapphire is suspended by wires in a vacuum chamber and cooled to near absolute zero....

The first accurate position measurement induces via the uncertainty principle a large momentum spread. For the same reason a collection of particles with different momenta will quickly drift apart, this induced spread in the bar's momentum soon results in a spread in the bar's position. Momentum just happens to be an attribute whose uncertainty feeds back into the position attribute. Braginski calls such a situation – where accurate measurement of one attribute is spoiled by the back-reaction of the Heisenberg spread in its conjugate attribute – <u>a</u> <u>quantum demolition measurement</u>. [37]

The function of the Gravito-Acoustic Oscillation and Amplification Chamber (Grand Gallery) was to oscillate and cohere, thereby amplifying, the electromagnetic, acoustic, and inertial harmonics of the base systems and to tune them to the target to establish a resonant and cohered standing wave within the target disrupting its nuclei and destroying it. The primary function of the Chamber is to structure the scalar potential of the target in the interaction region.

The electromagnetic resonances are coupled by two means: (1) the piezoelectric moment properties of the artificial crystals in each of its twenty-seven gravito-acoustic arrays, and (2) the optical resonances of the crystals themselves.

The piezoelectric properties of the crystals are stressed by the endothermic plasma state of the hydrogen plasma, as electrons are excited they emit photons, striking the crystals and causing similar nuclear resonances within each crystal, giving off the resonant

---

[37] Nick Herbert, *Quantum Reality: Beyond the New Physics, an Excursion into Metaphysics and the Meaning of Reality* (New York: Anchor Books, 1985), pp. 132-133, italicized emphasis added, underlined emphasis in the original.

optical frequencies of each crystal. This pulsing occurs in resonance to the Schumann resonance of the earth. The crystals are also pulsed by the optical and acoustic energy being amplified in the chamber due to its structure as a typical acoustic amplification chamber, and due to the polished surfaces of the limestone sides, floor, and ceiling of the chamber. Thus the crystal arrays are under constant stress from these three sources. Damping mechanisms attached to each of the arrays and each individual crystal would allow the chamber to be tuned to the electromagnetic, acoustic, and gravitational properties of the target.

The gravito-acoustic properties are accessed via two means: (1) the quantum scattering effects of each of the crystals, and (2) the difference in height from the center of the earth and corresponding slight variation in the acceleration of gravity at the varying heights of each of the crystal arrays.

Once the scalar potential of the target has been structured in the Chamber, its gravito-acoustic and electromagnetic output is conveyed to the Amplification Chamber for fine tuning and thence to the King's Chamber.

(a) The Twenty-Seven Gravito-Acoustic Resonator Arrays

There are twenty-seven banks of "gravito-acoustic resonator arrays" in the Grand Gallery (corresponding to the twenty-six dimensions of superstring theory?). The twenty-seventh bank would constitute the octave of the first bank, which is the fundamental.

These twenty-seven banks correspond to Sitchin's "magic stones". It is therefore postulated that the names in Sitchin's ancient Babylonian texts refer not to individual stones, but to the *arrays* of stones in each bank. Their names, now of course corrupted in transmission, perhaps originally designated the harmonic and physical action or energy to be oscillated.[38]

---

[38] That is, the names of the arrays preserved in paleographic texts cited by Sitchin would be the names ascribed to the functions of the arrays by less

These resonators were artificial crystals – perhaps in liquid form - constructed to be resonant to the corresponding gravity or pilot wave of any quantum interaction and any configuration of the base systems. Given the peculiar properties of sapphire already encountered, perhaps at least two crystals of each array, the bottom crystal and the top, were corundum, i.e., sapphire (bottom) and ruby(top). The stones in each array in between these two poles may have oscillated other colors of the visible light spectrum, since Sitchin's texts indicate that the interior of the chamber was once bathed in "a rainbow of colors." Given this statement, one has a basis on which to speculate about what other crystals might have been included in these arrays. The inclusion of ruby and sapphire (possibly in ϕ "Dark Crystal" form, cf. below, section 3, "the ϕ Crystals') poses an interesting problem touching both upon the principles of the unified paleophysics and upon the quest for unification in modern physics. The specific gravity of ruby and sapphire is the same since both are the same chemical compound. Yet, each is optically resonant to opposite ends of the spectrum indicating a connection between crystals, optics, electromagnetism, and gravity. (Investigation of this connection would lie in a reconsideration of known geometric lattice properties and chemical properties of crystals, or in the discovery of previously unknown properties based upon avenues of research suggested by paleophysical texts.)

(b) The ϕ Crystals and Crystals in Occult Tradition

There is a long tradition within occult and esoteric literature that associates certain crystals and gemstones with various zodiacal constellations. Viewed from the standpoint of a "paleophysical hermenuetic," these traditions may well represent some preservation of that sophisticated paleophysics that was engineered

---

sophisticated legacy civilizations attempting to understanding the physics principles and functions of each of the arrays.

in the Giza Death Star. Of primary interest are the traditions that associate sapphires with Giza.

> According to lore, The Book of the Angel Raziel was inscribed on a sapphire stone – the stone of destiny – and was once in the possession of Thoth/Enoch who, it is said, gave it out as his own work, i.e., The Book of Enoch/Thoth. IN The legends of the Jews from Primitive Times we learn that, like the Key of Life, Adam gave the stone to Seth, who gave it to Enoch, who passed it to noah, where it develops that Noah learned how to build the Ark by pouring over The Book.[39]

This "Destiny Stone" was maintained, "along with the other power tools," in the Great Pyramid.40 In a Hebrew legend, Abraham and Sarah discovered the rumored "Hall of Records" beneath the Sphinx. In it they found Thoth perfectly preserved and the prized "Emerald Tablets." William Henry speculated that these tablets held some key for "opening gateways to other parts of the universe."

> Let me offer an explanation for why this may be possible. Thoth's Emerald Tablets are associated with a shape Charles Hinton called the tesseract. Popular women's magazines of the late 1890's featured articles and advertisements featuring this curious cube.

---

[39] William Henry, *One Foot in Atlantis: The Secret Occult History of World War II and Its Impact on New Age Politics* (Anchorage, Alaska: Earthpulse Press, 1998), p. 143, emphasis added. Henry cites Robert Graves and Raphael Patai, *Hebrew Myths* (New York: Anchor Books, 1964), p. 113. It is worth mentioning that the psychic Edgar Cayce in some of his visions of Atlantean technology speaks of certain crystals in connection with gravity and destruction: "...(I)n Atlantean land at time of development of electrical forces that dealt with transportation of craft from place to place, photographing at a distance, overcoming gravity itself, preparation of the crystal, the terrible mighty crystal; much of this brought destruction."(519-1, February 20, 1934, cited in David Hatcher Childress, *Technology of the Gods: The Incredibale Science of the Ancients* [Kempton, Illinois: Adventures Unlimited Press, 2000], p. 296).

[40] Ibid., p. 182, citing Rene Guenon, *Fundamental Symbols: The Universal Language of Sacred Science* (Cambridge: Quinta Essentia, 1995), p. 121.

> The tesseract is a three-dimensional "shadow" of a four dimensional hypercube – a figure having a fourth dimension at right angles to the three with which we are familiar.[41]

Henry speculates on what this sapphire "destiny stone" may have done. Referring to the angels in the Biblical Jacob's vision of the Ladder, he states

> Are we to learn from Jacob's experience that:
> * The angels have the god-like ability to take the quantum equations of the three great German philosophers Einstein, Heisenberg, and Plank (sic.) and dissolve solid, molecular-atomic matter into cluster-waves of information and reassemble these waves into alternate forms? ...
> * Acquiring the (S)tone (sic.) of Destiny is a prerequisite for making this pact?
> I tend to think this (S)tone is the critical power tool.[42]

Henry is correct. The sapphire "destiny stone" is one of the most important components of the now-missing interior of the great weapon, the Giza Death Star, but *not* for any occult reason.

If one digs underneath the name "destiny stone" to the possible paleophysics the name is trying to convey or preserve, one comes on a rather interesting set of associations. "Destiny," in occult terms is linked with time, with the astrological signs of the zodiac. In more modern terms, it is linked to the geometric configuration of the planets and constellations, i.e., to the "gravitational harmonics" of space itself, and therefore to the planetary positions, masses, angular momenta and so on. "Destiny", in other words, is the legacy civilizations' shorthand for the paleophysical knowledge of the "interconnectedness between time, gravitational energy, acoustics, and geometry." Is there a link between gravity, acoustics, and sapphires explainable in terms of modern physical mechanics?

Indeed there is.

---

[41] Ibid., pp. 182-183.
[42] Ibid., p. 181.

If sapphire can be used to detect gravitational energy acoustically, then it is reasonable to conclude that one fundamental principle of the ancient paleophysics is radically different than our own: gravitational energy is not constant but variable as a function of geometry and harmonics, for the acoustic and harmonic connection to gravity is well-attested in the occult and esoteric literature.

> Certain Arab sources contain curious tales about the manner in which the pyramids of Egypt were erected. According to one, the stones were wrapped in papyrus and then struck with a rod by a priest. Thus they became completely weightless and moved through the air for about 50 meters. Then the hierophant repeated the procedure until the stone reached the pyramid and was put in place....
>
> Babylonian tablets affirm that sound could lift stones. The Bible speaks of Jericho and what sound waves did to its walls. Coptic writings relate the process by which blocks for the pyramids were elevated by the sound of chanting.[43]

Tales abound from Tibet where monks are reputed to be able to levitate stones or themselves by beating large drums. Moreover, "the use of horns and drums to acoustically levitate (sic.) something has been studied by NASA, and it is interesting to compare a modern stereo speaker cone with photos and diagrams of flying saucers."[44] If one were to use such sapphire or other crystals to couple acoustic, electromagnetic, nuclear and gravitational energy in a weapon, however, such crystals must have had some unusual properties not found in natural crystals.

We now enter an area of sheer speculation. Such crystals may have been artificially engineered not so much to refract light but to "capture" or absorb light via peculiar lattice properties. Such properties would give these crystals some very peculiar characteristics. I call these artificial crystals "ϕ" or "black crystals."

---

[43] Andrew Tomas, *We are not the First* (London: Souvenir Press, 1971), cited in David Hatcher Childress, *Technology of the Gods: The Incredible Sciences of the Ancients* (Kempton, Illinois: Adventures Unlimited Press, 2000), p. 160.

[44] Childress, op. cit., p. 162.

Such "phi" or "Black" Crystals would palpably resemble black holes and superconductors since such "phi Crystals" would also be analogous to a super-conductor, "imprisoning" electromagnetic energy by rotating electromagnetic fields inside of it by dint of its peculiar refractive and lattice properties. This would set up a field in the vicinity of the Pyramid that would literally "pull" or "tug" at anything on the surface or air space around it, exactly what was recorded by Sitchin's texts.

Three alternative hypotheses suggest themselves for the theoretical basis of such Dark Crystals:

- Black Crystals would absorb both EM, acoustic, and gravitational waves to the point that the crystal would exhibit mass and time dilation effects in local space and local time. This is consonant with the general hypothesis of the ancient paleophysics as having time as its primary differential.

- It is possible that the "Black Crystal" component of the gravito-acoustic arrays, if such a component was incorporated in them, would have had a refractive index of 1.61818 or a Fibonacci index. This would seem to imply some form of *liquid* crystals. There the energy is transcribed as acoustical or gravitic output. It is to be noted that ϕ is found in the Giza Death Star. As ϕ is the basis for the Fibonacci sequence and vorticular fluid mechanics, these crystals would have a kind of "Vorticular Refractive Index" and would thus seem to take on some properties of superconductors as well.

With respect to this "Fibonacci-vorticular" refractive index, it is to be noted that topaz and tourmaline offer good candidates not only for inclusion in the gravito-acoustic arrays in their natural states, but also for articificial modification into such ϕ crystals, since their refractive indices are already close approximations of the vorticular refraction index.

	Natural Crystal	
	Topaz	*Tourmaline*
Mohs Hardness	8	7-7.5
Specific Gravity	3.4-3.6	3.0-3.3
Refractive Index	1.61-1.63	1.62-1.65

(c) Hypothesized General Properties

The φ Crystals might thus have had the following general properties:

- Acoustic and Gravitational resonance accessed via piezoelectric stimulation of coherent phonon emission.
- Isotopic forms of one or more elements of their chemical molecular structure.
- A vorticular refractive index of 1.61818… .
- Possible hollow structure as Helmholtz resonators, or alternatively, liquid structure.
- Spherical structure, and possibly hollow in order to circumscribe some Platonic solid in crystal form.

(d) Required Sciences and Technologies

φ-Crystals, and indeed the Pyramid itself, may have required the knowledge of quantum computing in order for the crystal structures to be appropriately modeled and grown through the techniques of nanotechnology or some other method not now known. The growth of φ-Crystals for the fundamenal and octave of each array, the φ-Sapphires and φ-Rubies, would require the engineering of a lattice structure at the molecular level not found in nature in order to rotate light within the crystal. This would require that the crystals be assembled from the molecular level upwards in order to acquire the extremely close tolerances required in the lattice structure as will as the extremely close tolerances of the surface of each crystal to allow for minimum scattering of EM energy striking the surface of the crystals in order to transfer as much acoustic energy to the lattice structure itself. The function of the lattice structure is therefore apparent: It also serves as an amplification device for the acoustic energy down to the molecular, atomic, and quantum levels.

The connection between gravity and electromagnetism is assumed throughout the unified quaternion physics of Maxwell. The connection between gravity and acoustics, however, is a fundamental supposition of the paleophysics examined in The Giza Death Star, with its emphasis on harmonics.

Besides being well documented in esoteric traditions, this connection is currently a matter of serious scientific investigation. "Acoustic levitation" is the technique used to levitate an object by high-powered sounds, being used and proposed as a means of fueling targets in thermal nuclear reactions.[45]

The use of piezoelectric effect in the Gravito-acoustic crystal arrays of the Chamber are explainable in terms of contemporary theory.

> One example of "physical" acoustic study is about phonons, or the quanta of mechanic energy in a crystal lattice.... To my knowledge, crystal acoustics, as the acoustical study of phonons is called, uses excitation methods that are not "useful" in (a) conventional sense.... A crystal acoustician expects to introduce synchronized or coherent phonons by mechanical or photonic means.[46]

Several points emerge from careful consideration of Dr. Liu's remarks:

---

[45] Y. Liu, "Acoustics, an Unofficial Introduction, <www.stemnet.nf.ca/~yliu/acoustics.html>, p. 5.

[46] Ibid., p. 5, emphasis added.

- The conventional "non-usefulness" strongly suggests that a military application is being sought. (This interpretation is further strengthened in the remarks cited below.)
- The production of cohered phonons is obtained by regular photon stressing of a crystal lattice, exactly what is posited for the chamber from
  - The excitation of endothermic hydrogen plasma electrons and
  - The regular pulsing from the Schumann resonance.
- The notion of cohered phonon production corroborates the hypothesis suggested in The Giza Death Star. The piezoelectric effect is being employed to access the potential gravito-acoustic energy.

The connection between acoustics and gravity[47] is stressed by Dr. Liu as follows (the numbered points do not occur in his text, but are provided in order to facilitate the commentary that follows):

(1) Acoustic levitation has many advantages including its high controllability and good manipulation capabilities.

(2) Acoustic levitation was first experimentally verified in the 40s, but received intensive study only since Dr. T.G. Wang proposed its application in space material processing in 1979.

(3) JPL (the Jet Propulsion Laboratory) undoubtedly takes the lead in such studies. Yale University, MIT, Intersonics, Westinghouse, General Electrics (sic.), Marshal Space Center, Lawrence-Livermore Laboratories, and Bjorksfen Research Labs have all since taken their share...

(4) Acoustic levitation takes two main flavours: low-frequency levitation uses usual diaphragm loudspeakers to generate mostly audible sound (4oo Hz to 2Hz), while ultrasonic levitation uses piezoelectric transducers (frequency range 30 kHz to several megahertz). According to the formation of the field, there are uniaxial, triaxial, and focused field technique.[48]

The following implications occur with each enumerated point:

---

[47] It bears reapeating that noted electro-gravitics research physicist Thomas Townsend Brown also spent time researching the acoustic properties of rocks.

[48] Ibid., p. 6.

- The phenomenon is targetable, i.e., targeting occurs through harmonic interferometry, tending to confirm the hypothesis being advanced here.
- It is used to engineer materials not possible in terrestrial gravity.
- The involvement of the Jet Propulsion Laboratory, the Marshal Space Center, Lawrence-Livermore Laboratory as well as two major defense contractors (General Electric and Westinghouse), and two academic institutions with strong defense and intelligence connections suggests the militarization of the phenomenon for:
  - Kontrabaric (anti-gravity) Propulsion (NASA)
  - Materials engineering
  - Weaponization.

So one may be permitted to hypothesize that the missing components, the "magic stones" or "crystals" of Sitchin's texts, were some very sophisticated gravito-acoustic resonator assembly comprised of artificially engineered crystals designed to resonate to the harmonics of gravity.[49]   And another word is necessary. Recall that in Sitchin's texts the "Destiny" stone emitted a red radiance that Ninurta recorded as having a "strong power" that was used "to grab to pull me, with a tracking which kills to seize me."[50] Such a function would be fulfilled by sapphires or *rubies*.

And there is something else that must be noted. In Dunn's version of the machine hypothesis, the internal temperatures of the Pyramid would have been great, but not so great that the granite would have melted, as indicated by the melted Coffer and scorched limestone in the Grand Gallery. Cold fusion can take place at room temperatures.   Perhaps the Pyramid's more or less constant

---

[49] Needless to say, our current physics does not yet have an experimentally validated and fully developed knowledge of the harmonics of gravity. Braginski's experiment may be the first faltering steps in that direction.

[50] Zehcariah Sitchin, op. cit., p. 168. Cf. chapter two, p. 31.

temperature was used to maintain the hydrogen in a cold fusion state, but not to produce energy as much as to create the necessary analog of the sun and to duplicate or manipulate the gravitational effects to have been measured, resonated, and amplified by these crystalline arrays. The reason for the sharp upward incline of the Grand Gallery would thus serve *two* functions instead of only one. Not only was the design *acoustically* necessary as an amplification chamber, but gravitically necessary for accurate measures at different harmonically determined lengths from the center of the earth's mass. If this is so, then it stands to reason that at least some of these crystals were artificial sapphires and rubies. Needless to say, the discovery of any such crystalline artifacts of precise construction, resonance and appropriate age would constitute an extremely important archaeological find and a corroboration of this hypothesis.

(2) The Coffer:

Sitchin's texts also indicate components are missing from the inside of the Coffer. Perhaps, in addition to cohering the microwave input as an optical cavity, there were further devices inside of it that aided in the guidance or phasing of the output beam to couple it with the impulse discharge of the Pyramid-coil itself.

(3) The Subterranean Chamber(s):

Devices may have been placed here to "prime" the Pyramid, as Dunn suggests. Moreover, it is possible that some sort of instrumentation and machinery was placed here in order to "tune" and target the weapon properly and accurately.

(4) The Other Two Large Pyramids, The "Causeways", and the "Temples"

As Dunn indicates, the other structures at Giza may have been used to "prime" the Pyramid, and perhaps to aim it. It may be significant that the "causeways" leading to the three large pyramids, and the "temples" may have housed the necessary machinery to acquire a target and "aim" the weapon itself.

G. A Phase Conjugate Howitzer

So, what kind of weapon, after all, was the Great Pyramid, and how did it work?

It was an extremely sophisticated "phase conjugate mirror" *and* "phase conjugate howitzer" designed to collect, amplify, and cohere the acoustic, electromagnetic, nuclear, and "aetheric" or gravitational energy of sub-quantum local (i.e. terrestrial, solar, and galactic) space-time to a target in such a fashion that each of these several different types of energy arrived at the target at the same time and *exactly in phase, harmonically, with the target.* The "aetheric" energy, since it is the energy of non-local reality, functioned as the carrier wave or beam for the other forms of energy, guiding them to the target through hyperspace. With proper "tuning" of the weapon, any target, anywhere on earth or nearby space, could be selected. This superluminal carrier wave accelerated the electromagnetic, acoustic, and gravitational energy to the target.

Once the target had been selected and its harmonics were known, the weapon would have been tuned to hit it by properly tuning and configuring its missing internal components in conjunction with the other structures. Depending on what effect was desired in the target, one of two things could be done. A point in local space-time near or around the target would have been pulsed, releasing a violent discharge of electro-acoustical energy to the target. Accelerating the acoustical wave via this aetheric and electromagnetic discharge, a violent acoustic cavitation would be induced within the nuclei of its atoms until the target consumed itself in a violent nuclear reaction. And because of the nature of the weapon, *any material would do to set off that reaction. Wood, steel, or plastic would have been just as violently blown apart as uranium 235 or Plutonium.* Alternatively, the target could simply have been "slowly cooked", ionizing it altogether, a result no less violent, but considerably "cleaner". In either case, the pilot wave carries the electromagnetic and acoustic energy effortlessly into the nucleus, where those energies are accelerated to induce the cavitation.

The horror of such a weapon cannot be imagined. There would have been no radar warnings of incoming bombs, no tracks of blinking lights on oscilloscopes to watch. There would be little change in weather or atmospheric conditions to warn the victims of impending doom - save enormous electrical discharges paling the largest thunderstorms - unless its commanders wished to give it. There would be no infrared or other electromagnetic signature until it was almost too late. There would have been no moon-sized objects appearing in the sky over the target blasting it apart with enormous lasers; no saucers with gaping maws charging up capacitors and firing bolts of plasma at some ancient Empire State Building. And most importantly, after its use – always deadly, and always efficient – there would have been no more target, for it would have become the nuclear fuel, the critical mass, of its own consumption and ripped itself apart in a nuclear paroxysm of such violence that we can only imagine. Within certain limits, such a weapon could be calibrated to produce now a smaller, and now a much larger, blast, depending on how wide an area destruction was sought.

Lightening, falling from a clear blue sky, and a column of roaring smoke and fire on which a million deaths rides.

IX.
Who Built It?

"We may well discover that we are the Martians."
Richard C. Hoagland

A. Specialized Knowledge

One thing that is persistent in almost any ancient mythological text or tradition is the universal belief in "gods" each possessed of specific and precisely delineated "powers." From Australia to North and South America to the Near and Far East, the "gods" are gods "of the air", "of the sea", "of the spirits," "of oak" or "of trees" or "of light" and so on. There are gods of wisdom, gods of rivers, forests, winds, seas, oceans, fire, cold and virtually every other physical process or object we can think of. The normal response is to dismiss these traditions as either religiously or morally backward examples of primitive polytheism or pantheism, or the scientific illiteracy of primitive and backward cultures.

But I believe that there is nothing about our most ancient myths and traditions that is as it seems on the surface. Indeed, if there once existed a paleoancient Very High Civilization, as I believe the physics of the Great Pyramid demonstrates, then almost nothing about human history from the period of the demise of that Civilization to the appearance of the ancient civilizations of classical history can be taken for granted. And the poly- or pantheistic mythologies are no exception.

That being said, what do such myths in fact preserve from that paleoancient Very high Civilization? The answer may be discovered by extending the principle of paleographic interpretation of ancient texts that were employed in our examination of paleophysics and the archaeological evidence for a sophisticated ancient technology. If there was once a sophisticated "paleophysics", then one may assume that this paleophysics was as equally specialized into various sub-disciplines as our own, if not

282

more so. And by extension of that idea, one may assume that all of its science, art, and religion were similarly sophisticated, and compartmentalized.

Specialists would therefore have existed within each department of knowledge. A "wisdom-god" such as the Egyptian god Thoth might very well have been nothing more than a "minister of the department of research and development." An "ocean-god" a head of oceanography, and so on. The myths, after all, describe beings either superintending physical processes, specialized (meaning perhaps theoretical) knowledge, or human affairs, and sometimes mixtures of the three.

And this suggests a putative history, albeit a simplistic one, as to why the collective human memory of that period should have gone from humans in an advanced society, to "gods." The first step would have been taken in the period immediately after the demise of the paleoancient Very High Civilization. The men who ran, coordinated, and ultimately presided over the demise of the Civilization may have been referred to as being "men like as gods". The simile is remarkably preserved in the story of the fall of man in Genesis, for example.

The second step, as humanity slid further into backwardness as the infrastructure of the old civilization collapsed, would have been simply to transform the simile into a metaphor: "Men were gods." And the final step would be simply to drop the metaphor altogether: "There were once gods." The historical progression was thus the converse of the three-step program of French sociologist Auguste Comte. Comte, looking at the historical progress of man from the classical to the modern age, characterized the movement in terms of three broad epochs defined by the prevailing method employed to view the universe. These stages were the "religious, metaphysical, and scientific" ages. So the converse broad historical outline seems to have occurred some time in the remote past of human antiquity, in a movement from the scientific, to the metaphysical, to the religious age. This seems to be vindicated by the fact that here and there, in certain texts such as the *Hermetica*, the Hindu Epics, or the Sanskrit "vimana" texts, the outlines of vast technological and scientific sophistication are clearly preserved in spite of an evident overgrowth of the other two ages. But there is another aspect of those ancient myths and traditions that must also be accounted for.

B. *Long Average Life Spans*

Virtually every religious tradition from the ancient Near East has some tradition that mankind once lived much longer on average than he does now. The Babylonian texts, for example, attribute life spans of hundreds of thousands of years to man or the "gods", and the Old Testament, as is well known, also ascribes much longer life spans to humans. In some cases, life-spans of almost a millenium are reported. Typically, the response of mainstream research is to regard these reports as fanciful, as some kind of allegorical device to record the power of the gods or the venerable wisdom of ancestral forefathers.

But what if these reports, too, are the legacy of some paleoancient Very High Civilization? What if they in fact are true? It takes no great imagination to see how this might be possible. Already our own scientists are talking about the potentials of nano-technology[1] to repair the human body at the cellular level, cell by individual cell. Others refer to the stunning advances in genetics and the new and extremely powerful genetically based drugs and therapies for battling – and probably curing – diseases such as AIDS or cancer or even diabetes. Still others speak of "combinational" technologies that combine biological and mechanical techniques to open up a whole new vista for prosthetics. And just around the corner, one can envision the culturing and growth of genetically compatible organs for transplant when old organs "wear out." Man's technological quest

---

[1] i.e., the ability to manipulate and engineer reality at a molecular and atomic level by the construction of very tiny molecular-sized machines pre-programmed to perform certain functions, like tiny robots. Already scientists at AT&T have constructed the first artificial *atom.*

for virtual immortality appears poised on the cusp of stunning achievements.

So it is not difficult to imagine that our paleoancient Very High Civilization might have possessed the end results of technologies we are only just beginning to develop. And the implications of their possession of such technologies would have had a profound influence on the shape of their society and its cultural values.

(1) Rapid Scientific and Technological Progress

Obviously, the first important area where greatly expanded average life spans – in the hundreds or perhaps thousands of years – would have had a tremendous impact would be in the rapid accumulation and growth of knowledge. Once a civilization had mastered the techniques of nanotechnology and advanced genetic pharmacology and therapy, life spans would have increased, and with them, the accumulation of knowledge by individual members in a society would have grown exponentially. No longer would education have had the same "life and death" importance that it has for us.

Consider that every seventy years or so, all of humanity has to recycle all of its accumulated knowledge and pass it on to succeeding generations, and do so in a way that allows continued scientific and technological progress. Accordingly, our advancement in science and technology has proceeded at what amounts to a cultural snail's pace. Our most famous scientists, for example, are known for one or two major discoveries or accomplishments in their lives. Then, they die.

Consider also that with much longer life spans, scientists would perhaps be famous for vast, encyclopedic accumulations of discoveries. A culture and society in such a position would therefore be a society awash in the luxury of the leisure of being able to absorb such discoveries and progress within a single generation, and to press forward from them dramatically. It would be a society and culture whose knowledge is simultaneously very specialized and compartmentalized, much like our own. But unlike our own, the scientific breadth and depth of knowledge available to the non-specialist would be exponentially larger. Scientific literacy would be as much a staple to it as political, economic, or artistic literacy are to our own.

(2) Lower Population Density

With longer life spans, however, comes a cultural and perhaps even evolutionary consequence. On the one side, the human libido being what it is, one may speculate that strict measures would have been implemented to control population. Of course, if a society were as advanced as our paleoancient Very High Civilization, then this problem might not be all that serious. For example, as we have seen it doubtless relied on entirely different understandings of energy. It would not have been necessary for it to deplete the earth of various resources in order to fuel its energy needs, which, in any case, were probably too large for any amount of merely terrestrial resources to meet. Secondly, a such a highly advanced technological society would also imply a commensurately greater agricultural and food technology.

But all that being said, however, one is still left with human urges and libido, and so population density would have been a problem that it would have dealt with. What such measures would have been established one can only guess at. Perhaps some sort of program of enforced family sizes such as exists in modern China would have been mandated. We simply do not know.

But there is also another possibility at the opposite end of the spectrum that should not be overlooked. With greatly expanded life spans, the biological necessity for procreation might have produced some sort of evolutionary response within the human species. In short, humanity might have suffered a commensurate loss of sex drive.

Either way, what all this adds up to is that the paleoancient Very High Civilization was a society with a population density that was probably much less than our own. Large cities there probably were, but with the type of science and technology available to it, the need for them would have been correspondingly diminished.

(3) Its Global Extent and the Absence of a "Third World"

Its population density was probably therefore not only less than our own, but far more uniformly distributed. The distinction between rural and urban populations and classes would probably not have been as great nor as perceptible. And needless to say, the type of technology and science implied also means that there would have been no geopolitical distinction between the First, Second, or Third Worlds. There would have been but First World. This has a profound moral consequence that will be examined below.

C. *The Civilization Types*

The brilliant American theoretical physicist Michio Kaku reproduces an intriguing classification of civilizations based upon the types of physical forces they can access, control, and manipulate.

> It might be expected that only a more advanced civilization, with cast resources at its disposal, would have been capable of discovering the unified field theory. The astronomer Nikolai Kardashev, for example, has ranked advanced civilizations into three types: Type I civilizations, which control the resources of an entire planet; type II civilizations, which control the resources of a star; and type II civilizations, which control the resources of an entire galaxy.
>
> On this scale, technologically we are still on the threshold of achieving type I status. A true type I civilization would be able to perform feats far beyond the scope of present-day technology. For example, a type I civilization could not only predict the weather but control it. A type I civilization could make the Sahara desert bloom, harness the power of hurricanes for energy, change the course of rivers, harvest crops form the oceans... A type I civilization would be able to peer into the earth, predict of create earthquakes, and extract rare minerals and oil from inside the earth....

> The transition to a type II civilization, which can utilize and manipulate the power of the sun, may take several thousand years... a type II civilization could colonize the solar system and perhaps a few neighboring ones...and begin to build gigantic machines that can manipulate the greatest energy source in the solar system: the sun...
>
> The transition to a type III civilization, which can harness the resources of a galaxy, stretches our imagination to the limit.[2]

Based on these criteria, one can only conclude that the type of civilization that built the Great Pyramid was at the very least a type II, if not a type III, civilization, for the energy being manipulated in the Pyramid is precisely the energy of the sun, and even of the galaxy. That civilization was manifestly *not* the civilization of ancient Egypt, but something that must of necessity have predated it.

Kardashev's classification scheme and Kaku's interpretation of it is suggestive for another reason as well. Did the civilization that built the Great Pyramid have the capacity for interplanetary travel? The physics postulated for the Giza Death Star certainly seems to suggest that it did. But is there corroborative evidence?

I believe there is, and it is on Mars, where a Sphinx-like face and pyramidal structures peer up at us from the surface of our nearest neighbor. Richard C. Hoagland, who has done so much to advance the hypothesis of a once-civilized Mars and the physics of the society that existed there,[3] said it best. "We may well discover that *we* are the Martians."

This suggests another interesting method for verification of the weapon hypothesis, should we ever make a manned landing on the red planet. If the Mars-Cydonia monuments are the product of the same civilization that built the Great Pyramid, we should expect to find internal chambers to the D & M Pyramid and/or other pyramids on Mars to serve some similar machine-like or weapon function. Moreover, those structures may well have their internal

---

[2] Michio Kaku and Jennifer Thompson, *Beyond Einstein: the Cosmic Quest for the Theory of the Universe* (New York: Anchor Books, 1987), pp. 198-199.

[3] Cf. Richard C. Hoagland, *The Monuments of Mars: A City on the Edge of Forever.*

components still intact. That, of course, would be the ultimate verification, for it would tell us, once and for all, what type of machine and/or weapon the Great Pyramid was. Suffice it to say, however, that Hoagland and others have presented ample evidence that a sophisticated new physics of energy is at the very minimum being encoded in the monuments of Mars, if not manipulated and controlled. The evidence at Giza, of course, would indicate control and manipulation.

D. The Moral Condition

The most interesting aspect of this type of civilization and the most deep-seated impact of its scientific and technological prowess would have been on its moral condition. That it would have been a culture of great leisure seems rather obvious from the previous considerations.

However, there is a much more serious moral consequence stemming directly from a greatly expanded life span. The fourth century Christian father and bishop of Constantinople, St. John Chrysostom, once remarked that death was not only a punishment for sin but a remedy for it. And what he meant was simply that death was what set man apart from the angels and daemons because it cut off any further progress in evil.

His observation seems much more timely now than it did then, for a greatly expanded life span would mean one of two things for an individual in such a society. Either it would permit great moral progress in and toward the good, or great moral decay and "progress" in evil. Imagine an Albert Schweitzer or a Mother Teresa having thousands of years to do their work, or, conversely, an Adolf Hitler or a Joseph Stalin, and one has a picture of the moral condition that such a society might be in. The contrasts between good and evil both on an individual and on a societal basis would be very acute.

There are, however, other moral conditions to consider. A civilization possessed of very long average life spans would be, as has been said, a civilization of leisure. In a certain sense perhaps, it would be a civilization easily bored, with its constituent societies quick to take offense at minor insults or faults, or indeed, quick to find offense when none was intended. It would be, perhaps, a civilization of societies awash in petty jealousies and rivalries that could erupt, via their technological sophistication, with genocidal ferocity.

It is this aspect that I find so unsettling, beyond the implications of the Giza Death Star itself. Read a certain way, the ancient mythologies seem to us to be inexplicable. The gods of the myths more often than not display a pettiness, jealousy, and superficiality of moral vision that, even to our own culture's morally decayed standards, seems totally disproportionate to the power that they wield.

But viewed another way, of course, the "gods" were not gods at all, but men, and seen against the context of this hypothetical reconstruction of their civilization, their actions seem all too typical. And from what we have said throughout this book, we can see at least one monument on earth remains from a period when there was great progress in evil. The technology that made that civilization - and its maintenance - possible was twisted and turned against it in a breathtaking military application and a spasm of global destruction that – if one is to take seriously other ancient traditions – spilled over into environmental catastrophe.

It is chilling indeed to contemplate that the name of the supremely fallen one in the Judeo-Christian tradition is Lucifer, the "light-bearer," the preternaturally intelligent being whose special knowledge and "expertise" lies, after all, in electromagnetic phenomena.

King's Chamber.

The Giza Pyramid Complex from 4,000 Feet. This unique air photograph of the complex was taken before sunset at an elevation of 4,000 feet. The west slope of each pyramid is shown reflecting the light from the setting sun, and the south slope of each pyramid is in shadow. The south stepped "slope" of the Great Pyramid reveals a V-shaped depression in this photograph. The V-shaped depression, or hollowing-in, occurs in all four slopes of the Great Pyramid—a structural feature that does not appear on any other pyramid in the world.

The "Hollowing-In" Feature from Ground Level. This special feature of the Great Pyramid, extremely difficult to observe from ground level, did not escape the trained eye of Napoleon's artists, as may be seen in the etching above. A century was to pass before the structural engineer David Davidson would relate this "hollowing-in" feature to the three lengths of the year—solar, sideral, and anomalistic.

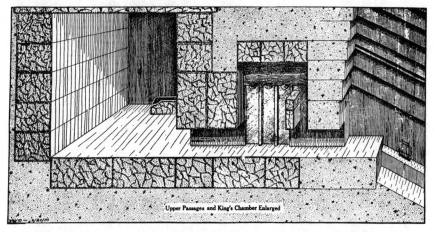

Upper Passages and King's Chamber. This cutaway drawing of the upper passages and King's Chamber is enlarged from scale for clarity.

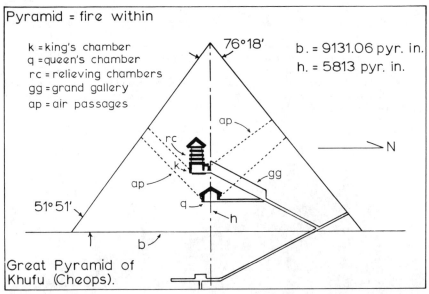

Pyramid = fire within

k = king's chamber
q = queen's chamber
rc = relieving chambers
gg = grand gallery
ap = air passages

76°18'

b. = 9131.06 pyr. in.
h. = 5813 pyr. in.

N

51°51'

Great Pyramid of Khufu (Cheops).

Professor Nelson suggests that if the coffer in the King's Chamber was filled with an aqueous solution of natron (N_AHCO_3, N_ACI, and $N_{A2}SO_4$), the salt-water itself would act as an effective conductor of electricity for the piezoelectric induction from the matt-finished walls of the King's Chamber. This, Professor Nelson points out, would make it unnecessary to line the coffer with metal — the salt itself is an effective conductor of electricity. Professor Nelson correctly points out that such a process would naturally produce poisonous chlorine gas which, somehow, would have been vented from the chamber.

The "stable" organic compounds in human blood are essentially the same as sea-water. A human candidate placed in this coffer during this process would experience a low voltage shock to his brain from the electrolyzed natron solution which Nelson says would have very good "health-restoring properties."

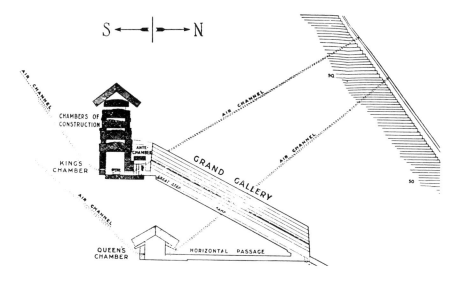

S ←—|—→ N

CHAMBERS OF
CONSTRUCTION

KINGS
CHAMBER

ANTE-
CHAMBER

AIR CHANNEL

AIR CHANNEL

AIR CHANNEL

AIR CHANNEL

GRAND GALLERY

GREAT STEP

RAMP

30

50

QUEEN'S
CHAMBER

HORIZONTAL PASSAGE

Cross Section Showing Air Channels.

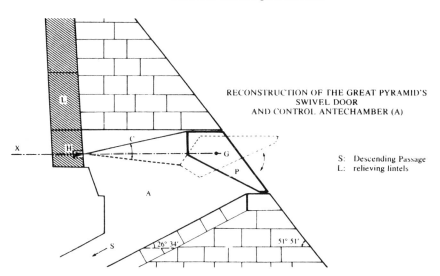

L

L

X

H

C

G

P

A

S

126° 34'

51° 51'

RECONSTRUCTION OF THE GREAT PYRAMID'S
SWIVEL DOOR
AND CONTROL ANTECHAMBER (A)

S: Descending Passage
L: relieving lintels

Reconstruction of the Great Pyramid's Swivel Door and Control Antechamber.

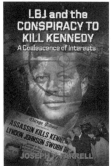

SECRETS OF THE UNIFIED FIELD
The Philadelphia Experiment, the Nazi Bell, and the Discarded Theory
by Joseph P. Farrell

Farrell examines the now discarded Unified Field Theory. American and German wartime scientists and engineers determined that, while the theory was incomplete, it could nevertheless be engineered. Chapters include: The Meanings of "Torsion"; Wringing an Aluminum Can; The Mistake in Unified Field Theories and Their Discarding by Contemporary Physics; Three Routes to the Doomsday Weapon: Quantum Potential, Torsion, and Vortices; Tesla's Meeting with FDR; Arnold Sommerfeld and Electromagnetic Radar Stealth; Electromagnetic Phase Conjugations, Phase Conjugate Mirrors, and Templates; The Unified Field Theory, the Torsion Tensor, and Igor Witkowski's Idea of the Plasma Focus; tons more.
340 pages. 6x9 Paperback. Illustrated. Bibliography. Index. $18.95. Code: SOUF

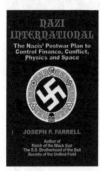

NAZI INTERNATIONAL
The Nazi's Postwar Plan to Control Finance, Conflict, Physics and Space
by Joseph P. Farrell

Beginning with prewar corporate partnerships in the USA, including some with the Bush family, he moves on to the surrender of Nazi Germany, and evacuation plans of the Germans. He then covers the vast, and still-little-known recreation of Nazi Germany in South America with help of Juan Peron, I.G. Farben and Martin Bormann. Farrell then covers Nazi Germany's penetration of the Muslim world including Wilhelm Voss and Otto Skorzeny in Gamel Abdul Nasser's Egypt before moving on to the development and control of new energy technologies including the Bariloche Fusion Project, Dr. Philo Farnsworth's Plasmator, and the work of Dr. Nikolai Kozyrev. Finally, Farrell discusses the Nazi desire to control space, and examines their connection with NASA, the esoteric meaning of NASA Mission Patches.
412 pages. 6x9 Paperback. Illustrated. References. $19.95. Code: NZIN

ARKTOS
The Polar Myth in Science, Symbolism & Nazi Survival
by Joscelyn Godwin

Explored are the many tales of an ancient race said to have lived in the Arctic regions, such as Thule and Hyperborea. Progressing onward, he looks at modern polar legends: including the survival of Hitler, German bases in Antarctica, UFOs, the hollow earth, and the hidden kingdoms of Agartha and Shambala. Chapters include: Prologue in Hyperborea; The Golden Age; The Northern Lights; The Arctic Homeland; The Aryan Myth; The Thule Society; The Black Order; The Hidden Lands; Agartha and the Polaires; Shambhala; The Hole at the Pole; Antarctica; more.
220 Pages. 6x9 Paperback. Illustrated. Bib. Index. $16.95. Code: ARK

MIND CONTROL, WORLD CONTROL
The Encyclopedia of Mind Control
by Jim Keith

Keith uncovers a surprising amount of information on the technology, experimentation and implementation of Mind Control technology. Various chapters in this shocking book are on early C.I.A. experiments such as Project Artichoke and Project RIC-EDOM, the methodology and technology of implants, Mind Control Assassins and Couriers, various famous "Mind Control" victims such as Sirhan Sirhan and Candy Jones. Also featured in this book are chapters on how Mind Control technology may be linked to some UFO activity and "UFO abductions.
256 Pages. 6x9 Paperback. Illustrated. $14.95. Code: MCWC

LOST CITIES & ANCIENT MYSTERIES OF THE SOUTHWEST
By David Hatcher Childress

Join David as he starts in northern Mexico and searches for the lost mines of the Aztecs. He continues north to west Texas, delving into the mysteries of Big Bend, including mysterious Phoenician tablets discovered there and the strange lights of Marfa. Then into New Mexico where he stumbles upon a hollow mountain with a billion dollars of gold bars hidden deep inside it! In Arizona he investigates tales of Egyptian catacombs in the Grand Canyon, cruises along the Devil's Highway, and tackles the century-old mystery of the Lost Dutchman mine. In Nevada and California Childress checks out the rumors of mummified giants and weird tunnels in Death Valley, plus he searches the Mohave Desert for the mysterious remains of ancient dwellers alongside lakes that dried up tens of thousands of years ago. It's a full-tilt blast down the back roads of the Southwest in search of the weird and wondrous mysteries of the past!

486 Pages. 6x9 Paperback. Illustrated. Bibliography. $19.95. Code: LCSW

TECHNOLOGY OF THE GODS
The Incredible Sciences of the Ancients
by David Hatcher Childress

Childress looks at the technology that was allegedly used in Atlantis and the theory that the Great Pyramid of Egypt was originally a gigantic power station. He examines tales of ancient flight and the technology that it involved; how the ancients used electricity; megalithic building techniques; the use of crystal lenses and the fire from the gods; evidence of various high tech weapons in the past, including atomic weapons; ancient metallurgy and heavy machinery; the role of modern inventors such as Nikola Tesla in bringing ancient technology back into modern use; impossible artifacts; and more.

356 PAGES. 6x9 PAPERBACK. ILLUSTRATED. BIBLIOGRAPHY. $16.95. CODE: TGOD

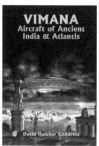

VIMANA AIRCRAFT OF ANCIENT INDIA & ATLANTIS
by David Hatcher Childress, introduction by Ivan T. Sanderson

In this incredible volume on ancient India, authentic Indian texts such as the *Ramayana* and the *Mahabharata* are used to prove that ancient aircraft were in use more than four thousand years ago. Included in this book is the entire Fourth Century BC manuscript *Vimaanika Shastra* by the ancient author Maharishi Bharadwaaja. Also included are chapters on Atlantean technology, the incredible Rama Empire of India and the devastating wars that destroyed it.

334 PAGES. 6x9 PAPERBACK. ILLUSTRATED. $15.95. CODE: VAA

LOST CONTINENTS & THE HOLLOW EARTH
I Remember Lemuria and the Shaver Mystery
by David Hatcher Childress & Richard Shaver

Shaver's rare 1948 book *I Remember Lemuria* is reprinted in its entirety, and the book is packed with illustrations from Ray Palmer's *Amazing Stories* magazine of the 1940s. Palmer and Shaver told of tunnels running through the earth—tunnels inhabited by the Deros and Teros, humanoids from an ancient spacefaring race that had inhabited the earth, eventually going underground, hundreds of thousands of years ago. Childress discusses the famous hollow earth books and delves deep into whatever reality may be behind the stories of tunnels in the earth. Operation High Jump to Antarctica in 1947 and Admiral Byrd's bizarre statements, tunnel systems in South America and Tibet, the underground world of Agartha, the belief of UFOs coming from the South Pole, more.

344 PAGES. 6x9 PAPERBACK. ILLUSTRATED. $16.95. CODE: LCHE

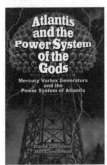

ATLANTIS & THE POWER SYSTEM OF THE GODS
by David Hatcher Childress and Bill Clendenon
Childress' fascinating analysis of Nikola Tesla's broadcast system in light of Edgar Cayce's "Terrible Crystal" and the obelisks of ancient Egypt and Ethiopia. Includes: Atlantis and its crystal power towers that broadcast energy; how these incredible power stations may still exist today; inventor Nikola Tesla's nearly identical system of power transmission; Mercury Proton Gyros and mercury vortex propulsion; more. Richly illustrated, and packed with evidence that Atlantis not only existed—it had a world-wide energy system more sophisticated than ours today.
246 PAGES. 6x9 PAPERBACK. ILLUSTRATED. $15.95. CODE: APSG

THE ANTI-GRAVITY HANDBOOK
edited by David Hatcher Childress
The new expanded compilation of material on Anti-Gravity, Free Energy, Flying Saucer Propulsion, UFOs, Suppressed Technology, NASA Cover-ups and more. Highly illustrated with patents, technical illustrations and photos. This revised and expanded edition has more material, including photos of Area 51, Nevada, the government's secret testing facility. This classic on weird science is back in a new format!
230 PAGES. 7x10 PAPERBACK. ILLUSTRATED. $16.95. CODE: AGH

ANTI–GRAVITY & THE WORLD GRID
Is the earth surrounded by an intricate electromagnetic grid network offering free energy? This compilation of material on ley lines and world power points contains chapters on the geography, mathematics, and light harmonics of the earth grid. Learn the purpose of ley lines and ancient megalithic structures located on the grid. Discover how the grid made the Philadelphia Experiment possible. Explore the Coral Castle and many other mysteries, including acoustic levitation, Tesla Shields and scalar wave weaponry. Browse through the section on anti-gravity patents, and research resources.
274 PAGES. 7x10 PAPERBACK. ILLUSTRATED. $14.95. CODE: AGW

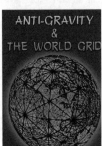
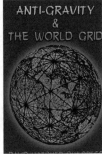

ANTI–GRAVITY & THE UNIFIED FIELD
edited by David Hatcher Childress
Is Einstein's Unified Field Theory the answer to all of our energy problems? Explored in this compilation of material is how gravity, electricity and magnetism manifest from a unified field around us. Why artificial gravity is possible; secrets of UFO propulsion; free energy; Nikola Tesla and anti-gravity airships of the 20s and 30s; flying saucers as superconducting whirls of plasma; anti-mass generators; vortex propulsion; suppressed technology; government cover-ups; gravitational pulse drive; spacecraft & more.
240 PAGES. 7x10 PAPERBACK. ILLUSTRATED. $14.95. CODE: AGU

THE TIME TRAVEL HANDBOOK
A Manual of Practical Teleportation & Time Travel
edited by David Hatcher Childress
The Time Travel Handbook takes the reader beyond the government experiments and deep into the uncharted territory of early time travellers such as Nikola Tesla and Guglielmo Marconi and their alleged time travel experiments, as well as the Wilson Brothers of EMI and their connection to the Philadelphia Experiment—the U.S. Navy's forays into invisibility, time travel, and teleportation. Childress looks into the claims of time travelling individuals, and investigates the unusual claim that the pyramids on Mars were built in the future and sent back in time. A highly visual, large format book, with patents, photos and schematics. Be the first on your block to build your own time travel device!
316 PAGES. 7x10 PAPERBACK. ILLUSTRATED. $16.95. CODE: TTH

MAPS OF THE ANCIENT SEA KINGS
Evidence of Advanced Civilization in the Ice Age
by Charles H. Hapgood

Charles Hapgood has found the evidence in the Piri Reis Map that shows Antarctica, the Hadji Ahmed map, the Oronteus Finaeus and other amazing maps. Hapgood concluded that these maps were made from more ancient maps from the various ancient archives around the world, now lost. Not only were these unknown people more advanced in mapmaking than any people prior to the 18th century, it appears they mapped all the continents. The Americas were mapped thousands of years before Columbus. Antarctica was mapped when its coasts were free of ice!

316 PAGES. 7x10 PAPERBACK. ILLUSTRATED. BIBLIOGRAPHY & INDEX. $19.95. CODE: MASK

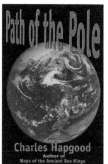

PATH OF THE POLE
Cataclysmic Pole Shift Geology
by Charles H. Hapgood

Maps of the Ancient Sea Kings author Hapgood's classic book *Path of the Pole* is back in print! Hapgood researched Antarctica, ancient maps and the geological record to conclude that the Earth's crust has slipped on the inner core many times in the past, changing the position of the pole. *Path of the Pole* discusses the various "pole shifts" in Earth's past, giving evidence for each one, and moves on to possible future pole shifts.

356 PAGES. 6x9 PAPERBACK. ILLUSTRATED. $16.95. CODE: POP

SECRETS OF THE HOLY LANCE
The Spear of Destiny in History & Legend
by Jerry E. Smith

Secrets of the Holy Lance traces the Spear from its possession by Constantine, Rome's first Christian Caesar, to Charlemagne's claim that with it he ruled the Holy Roman Empire by Divine Right, and on through two thousand years of kings and emperors, until it came within Hitler's grasp—and beyond! Did it rest for a while in Antarctic ice? Is it now hidden in Europe, awaiting the next person to claim its awesome power? Neither debunking nor worshiping, *Secrets of the Holy Lance* seeks to pierce the veil of myth and mystery around the Spear. Mere belief that it was infused with magic by virtue of its shedding the Savior's blood has made men kings. But what if it's more? What are "the powers it serves"?

312 PAGES. 6x9 PAPERBACK. ILLUSTRATED. BIBLIOGRAPHY. $16.95. CODE: SOHL

THE FANTASTIC INVENTIONS OF NIKOLA TESLA
by Nikola Tesla with additional material by
David Hatcher Childress

This book is a readable compendium of patents, diagrams, photos and explanations of the many incredible inventions of the originator of the modern era of electrification. In Tesla's own words are such topics as wireless transmission of power, death rays, and radio-controlled airships. In addition, rare material on a secret city built at a remote jungle site in South America by one of Tesla's students, Guglielmo Marconi. Marconi's secret group claims to have built flying saucers in the 1940s and to have gone to Mars in the early 1950s! Incredible photos of these Tesla craft are included. •His plan to transmit free electricity into the atmosphere. •How electrical devices would work using only small antennas. •Why unlimited power could be utilized anywhere on earth. •How radio and radar technology can be used as death-ray weapons in Star Wars.

342 PAGES. 6x9 PAPERBACK. ILLUSTRATED. $16.95. CODE: FINT

REICH OF THE BLACK SUN
Nazi Secret Weapons & the Cold War Allied Legend
by Joseph P. Farrell

Why were the Allies worried about an atom bomb attack by the Germans in 1944? Why did the Soviets threaten to use poison gas against the Germans? Why did Hitler in 1945 insist that holding Prague could win the war for the Third Reich? Why did US General George Patton's Third Army race for the Skoda works at Pilsen in Czechoslovakia instead of Berlin? Why did the US Army not test the uranium atom bomb it dropped on Hiroshima? Why did the Luftwaffe fly a non-stop round trip mission to within twenty miles of New York City in 1944? *Reich of the Black Sun* takes the reader on a scientific-historical journey in order to answer these questions. Arguing that Nazi Germany actually won the race for the atom bomb in late 1944,

352 PAGES. 6x9 PAPERBACK. ILLUSTRATED. BIBLIOGRAPHY. $16.95.
CODE: ROBS

THE GIZA DEATH STAR
The Paleophysics of the Great Pyramid & the Military Complex at Giza
by Joseph P. Farrell

Was the Giza complex part of a military installation over 10,000 years ago? Chapters include: An Archaeology of Mass Destruction, Thoth and Theories; The Machine Hypothesis; Pythagoras, Plato, Planck, and the Pyramid; The Weapon Hypothesis; Encoded Harmonics of the Planck Units in the Great Pyramid; High Freqquency Direct Current "Impulse" Technology; The Grand Gallery and its Crystals: Gravito-acoustic Resonators; The Other Two Large Pyramids; the "Causeways," and the "Temples"; A Phase Conjugate Howitzer; Evidence of the Use of Weapons of Mass Destruction in Ancient Times; more.

290 PAGES. 6x9 PAPERBACK. ILLUSTRATED. $16.95. CODE: GDS

THE GIZA DEATH STAR DEPLOYED
The Physics & Engineering of the Great Pyramid
by Joseph P. Farrell

Farrell expands on his thesis that the Great Pyramid was a maser, designed as a weapon and eventually deployed—with disastrous results to the solar system. Includes: Exploding Planets: A Brief History of the Exoteric and Esoteric Investigations of the Great Pyramid; No Machines, Please!; The Stargate Conspiracy; The Scalar Weapons; Message or Machine?; A Tesla Analysis of the Putative Physics and Engineering of the Giza Death Star; Cohering the Zero Point, Vacuum Energy, Flux: Feedback Loops and Tetrahedral Physics; and more.

290 PAGES. 6x9 PAPERBACK. ILLUSTRATED. $16.95. CODE: GDSD

THE GIZA DEATH STAR DESTROYED
The Ancient War For Future Science
by Joseph P. Farrell

Farrell moves on to events of the final days of the Giza Death Star and its awesome power. These final events, eventually leading up to the destruction of this giant machine, are dissected one by one, leading us to the eventual abandonment of the Giza Military Complex—an event that hurled civilization back into the Stone Age. Chapters include: The Mars-Earth Connection; The Lost "Root Races" and the Moral Reasons for the Flood; The Destruction of Krypton: The Electrodynamic Solar System, Exploding Planets and Ancient Wars; Turning the Stream of the Flood: the Origin of Secret Societies and Esoteric Traditions; The Quest to Recover Ancient Mega-Technology; Non-Equilibrium Paleophysics; Monatomic Paleophysics; Frequencies, Vortices and Mass Particles; "Acoustic" Intensity of Fields; The Pyramid of Crystals; tons more.

292 pages. 6x9 paperback. Illustrated. $16.95. Code: GDES

THE TESLA PAPERS
Nikola Tesla on Free Energy &
Wireless Transmission of Power
by Nikola Tesla, edited by David Hatcher Childress
David Hatcher Childress takes us into the incredible world of Nikola Tesla and his amazing inventions. Tesla's fantastic vision of the future, including wireless power, anti-gravity, free energy and highly advanced solar power. Also included are some of the papers, patents and material collected on Tesla at the Colorado Springs Tesla Symposiums, including papers on: •The Secret History of Wireless Transmission •Tesla and the Magnifying Transmitter •Design and Construction of a Half-Wave Tesla Coil •Electrostatics: A Key to Free Energy •Progress in Zero-Point Energy Research •Electromagnetic Energy from Antennas to Atoms •Tesla's Particle Beam Technology •Fundamental Excitatory Modes of the Earth-Ionosphere Cavity
325 PAGES. 8x10 PAPERBACK. ILLUSTRATED. $16.95. CODE: TTP

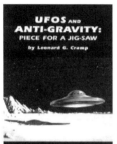

UFOS AND ANTI-GRAVITY
Piece For A Jig-Saw
by Leonard G. Cramp
Leonard G. Cramp's 1966 classic book on flying saucer propulsion and suppressed technology is a highly technical look at the UFO phenomena by a trained scientist. Cramp first introduces the idea of 'anti-gravity' and introduces us to the various theories of gravitation. He then examines the technology necessary to build a flying saucer and examines in great detail the technical aspects of such a craft. Cramp's book is a wealth of material and diagrams on flying saucers, anti-gravity, suppressed technology, G-fields and UFOs. Chapters include Crossroads of Aerodymanics, Aerodynamic Saucers, Limitations of Rocketry, Gravitation and the Ether, Gravitational Spaceships, G-Field Lift Effects, The Bi-Field Theory, VTOL and Hovercraft, Analysis of UFO photos, more.
388 PAGES. 6x9 PAPERBACK. ILLUSTRATED. $16.95. CODE: UAG

THE COSMIC MATRIX
Piece for a Jig-Saw, Part Two
by Leonard G. Cramp
Cramp examines anti-gravity effects and theorizes that this super-science used by the craft—described in detail in the book—can lift mankind into a new level of technology, transportation and understanding of the universe. The book takes a close look at gravity control, time travel, and the interlocking web of energy between all planets in our solar system with Leonard's unique technical diagrams. A fantastic voyage into the present and future!
364 PAGES. 6x9 PAPERBACK. ILLUSTRATED. BIBLIOGRAPHY. $16.00. CODE: CMX

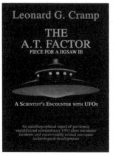

THE A.T. FACTOR
A Scientists Encounter with UFOs
by Leonard Cramp
British aerospace engineer Cramp began much of the scientific anti-gravity and UFO propulsion analysis back in 1955 with his landmark book *Space, Gravity & the Flying Saucer* (out-of-print and rare). In this final book, Cramp brings to a close his detailed and controversial study of UFOs and Anti-Gravity.
324 PAGES. 6x9 PAPERBACK. ILLUSTRATED. BIBLIOGRAPHY. INDEX. $16.95. CODE: ATF

THE FREE-ENERGY DEVICE HANDBOOK
A Compilation of Patents and Reports
by David Hatcher Childress

A large-format compilation of various patents, papers, descriptions and diagrams concerning free-energy devices and systems. *The Free-Energy Device Handbook* is a visual tool for experimenters and researchers into magnetic motors and other "over-unity" devices. With chapters on the Adams Motor, the Hans Coler Generator, cold fusion, superconductors, "N" machines, space-energy generators, Nikola Tesla, T. Townsend Brown, and the latest in free-energy devices. Packed with photos, technical diagrams, patents and fascinating information, this book belongs on every science shelf.

292 PAGES. 8x10 PAPERBACK. ILLUSTRATED. $16.95. CODE: FEH

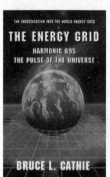

THE ENERGY GRID
Harmonic 695, The Pulse of the Universe
by Captain Bruce Cathie

This is the breakthrough book that explores the incredible potential of the Energy Grid and the Earth's Unified Field all around us. Cathie's first book, *Harmonic 33*, was published in 1968 when he was a commercial pilot in New Zealand. Since then, Captain Bruce Cathie has been the premier investigator into the amazing potential of the infinite energy that surrounds our planet every microsecond. Cathie investigates the Harmonics of Light and how the Energy Grid is created. In this amazing book are chapters on UFO Propulsion, Nikola Tesla, Unified Equations, the Mysterious Aerials, Pythagoras & the Grid, Nuclear Detonation and the Grid, Maps of the Ancients, an Australian Stonehenge examined, more.

255 PAGES. 6x9 TRADEPAPER. ILLUSTRATED. $15.95. CODE: TEG

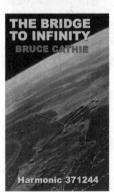

THE BRIDGE TO INFINITY
Harmonic 371244
by Captain Bruce Cathie

Cathie has popularized the concept that the earth is crisscrossed by an electromagnetic grid system that can be used for anti-gravity, free energy, levitation and more. The book includes a new analysis of the harmonic nature of reality, acoustic levitation, pyramid power, harmonic receiver towers and UFO propulsion. It concludes that today's scientists have at their command a fantastic store of knowledge with which to advance the welfare of the human race.

204 PAGES. 6x9 TRADEPAPER. ILLUSTRATED. $14.95. CODE: BTF

THE HARMONIC CONQUEST OF SPACE
by Captain Bruce Cathie

Chapters include: Mathematics of the World Grid; the Harmonics of Hiroshima and Nagasaki; Harmonic Transmission and Receiving; the Link Between Human Brain Waves; the Cavity Resonance between the Earth; the Ionosphere and Gravity; Edgar Cayce—the Harmonics of the Subconscious; Stonehenge; the Harmonics of the Moon; the Pyramids of Mars; Nikola Tesla's Electric Car; the Robert Adams Pulsed Electric Motor Generator; Harmonic Clues to the Unified Field; and more. Also included are tables showing the harmonic relations between the earth's magnetic field, the speed of light, and anti-gravity/gravity acceleration at different points on the earth's surface. New chapters in this edition on the giant stone spheres of Costa Rica, Atomic Tests and Volcanic Activity, and a chapter on Ayers Rock analysed with Stone Mountain, Georgia.

248 PAGES. 6x9. PAPERBACK. ILLUSTRATED. BIBLIOGRAPHY. $16.95. CODE: HCS

GRAVITATIONAL MANIPULATION OF DOMED CRAFT
UFO Propulsion Dynamics
by Paul E. Potter

Potter's precise and lavish illustrations allow the reader to enter directly into the realm of the advanced technological engineer and to understand, quite straightforwardly, the aliens' methods of energy manipulation: their methods of electrical power generation; how they purposely designed their craft to employ the kinds of energy dynamics that are exclusive to space (discoverable in our astrophysics) in order that their craft may generate both attractive and repulsive gravitational forces; their control over the mass-density matrix surrounding their craft enabling them to alter their physical dimensions and even manufacture their own frame of reference in respect to time. Includes a 16-page color insert.

624 pages. 7x10 Paperback. Illustrated. References. $24.00. Code: GMDC

TAPPING THE ZERO POINT ENERGY
Free Energy & Anti-Gravity in Today's Physics
by Moray B. King

King explains how free energy and anti-gravity are possible. The theories of the zero point energy maintain there are tremendous fluctuations of electrical field energy imbedded within the fabric of space. This book tells how, in the 1930s, inventor T. Henry Moray could produce a fifty kilowatt "free energy" machine; how an electrified plasma vortex creates anti-gravity; how the Pons/Fleischmann "cold fusion" experiment could produce tremendous heat without fusion; and how certain experiments might produce a gravitational anomaly.

180 PAGES. 5x8 PAPERBACK. ILLUSTRATED. $12.95. CODE: TAP

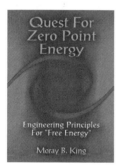

QUEST FOR ZERO-POINT ENERGY
Engineering Principles for "Free Energy"
by Moray B. King

King expands, with diagrams, on how free energy and anti-gravity are possible. The theories of zero point energy maintain there are tremendous fluctuations of electrical field energy embedded within the fabric of space. King explains the following topics: TFundamentals of a Zero-Point Energy Technology; Vacuum Energy Vortices; The Super Tube; Charge Clusters: The Basis of Zero-Point Energy Inventions; Vortex Filaments, Torsion Fields and the Zero-Point Energy; Transforming the Planet with a Zero-Point Energy Experiment; Dual Vortex Forms: The Key to a Large Zero-Point Energy Coherence. Packed with diagrams, patents and photos.

224 PAGES. 6x9 PAPERBACK. ILLUSTRATED. $14.95. CODE: QZPE

DARK MOON
Apollo and the Whistleblowers
by Mary Bennett and David Percy

Did you know a second craft was going to the Moon at the same time as Apollo 11? Do you know that potentially lethal radiation is prevalent throughout deep space? Do you know there are serious discrepancies in the account of the Apollo 13 'accident'? Did you know that 'live' color TV from the Moon was not actually live at all? Did you know that the Lunar Surface Camera had no viewfinder? Do you know that lighting was used in the Apollo photographs—yet no lighting equipment was taken to the Moon? All these questions, and more, are discussed in great detail by British researchers Bennett and Percy in *Dark Moon*, the definitive book (nearly 600 pages) on the possible faking of the Apollo Moon missions. Tons of NASA photos analyzed for possible deceptions.

568 PAGES. 6x9 PAPERBACK. ILLUSTRATED. BIBLIOGRAPHY. INDEX. $32.00. CODE: DMO

ORDER FORM

**10% Discount
When You Order
3 or More Items!**

One Adventure Place
P.O. Box 74
Kempton, Illinois 60946
United States of America
Tel.: 815-253-6390 • Fax: 815-253-6300
Email: auphq@frontiernet.net
http://www.adventuresunlimitedpress.com

ORDERING INSTRUCTIONS

✓ Remit by USD$ Check, Money Order or Credit Card

✓ Visa, Master Card, Discover & AmEx Accepted

✓ Paypal Payments Can Be Made To:

   info@wexclub.com

✓ Prices May Change Without Notice

✓ 10% Discount for 3 or more Items

SHIPPING CHARGES

United States

✓ Postal Book Rate { $4.00 First Item
50¢ Each Additional Item

✓ POSTAL BOOK RATE Cannot Be Tracked!

✓ Priority Mail { $5.00 First Item
$2.00 Each Additional Item

✓ UPS { $6.00 First Item
$1.50 Each Additional Item

NOTE: UPS Delivery Available to Mainland USA Only

Canada

✓ Postal Air Mail { $10.00 First Item
$2.50 Each Additional Item

✓ Personal Checks or Bank Drafts MUST BE

   US$ and Drawn on a US Bank

✓ Canadian Postal Money Orders OK

✓ Payment MUST BE US$

All Other Countries

✓ Sorry, No Surface Delivery!

✓ Postal Air Mail { $16.00 First Item
$6.00 Each Additional Item

✓ Checks and Money Orders MUST BE US$
   and Drawn on a US Bank or branch.

✓ Paypal Payments Can Be Made in US$ To:
   info@wexclub.com

SPECIAL NOTES

✓ RETAILERS: Standard Discounts Available

✓ BACKORDERS: We Backorder all Out-of-
   Stock Items Unless Otherwise Requested

✓ PRO FORMA INVOICES: Available on Request

ORDER ONLINE AT: www.adventuresunlimitedpress.com

Please check: ✓

☐ This is my first order   ☐ I have ordered before

Name			
Address			
City			
State/Province		Postal Code	
Country			
Phone day		Evening	
Fax		Email	

Item Code	Item Description	Qty	Total

Please check: ✓

☐ Postal-Surface

☐ Postal-Air Mail
   (Priority in USA)

☐ UPS
   (Mainland USA only)

☐ Visa/MasterCard/Discover/American Express

Subtotal ▶	
Less Discount-10% for 3 or more items ▶	
Balance ▶	
Illinois Residents 6.25% Sales Tax ▶	
Previous Credit ▶	
Shipping ▶	
Total (check/MO in USD$ only) ▶	

Card Number

Expiration Date

10% Discount When You Order 3 or More Items!